Experiment Earth

In recent years, experiments in geoengineering – intentionally manipulating the Earth's climate to reduce global warming – have become the focus of a vital debate about the intended and unintended consequences of innovation, raising profound social, political and ethical questions.

This book explores these issues through the lens of the research project SPICE (Stratospheric Particle Injection for Climate Engineering), one of the first major geoengineering studies worldwide which aims to put particles high into the atmosphere to cut the amount of sunlight reaching the Earth's surface. Drawing on three years of sociological research working with the scientists investigating the idea of geoengineering, the book examines how experiments become controversial and why many are calling for the scientific community and civil society to rethink how we govern emerging technologies. It illustrates broader dynamics that are highly relevant to wider debates on science and technology governance and the responsibilities of scientists to take better care of the futures they help bring about.

This book takes a critical stance on existing assumptions about ethical issues in science, giving students, researchers and the general reader interested in the place of science in contemporary society a compelling framework for future thinking and discussion.

Jack Stilgoe is a lecturer in the Department of Science and Technology Studies at University College London, UK.

'How should society react when the technological imagination seizes on the Earth itself as an experimental system? In this graceful critique of magical thinking, Stilgoe dissects the moves by which some came to see geoengineering as a project that not only can be done but must be done. An essential addition to the renewed debate on climate change, the book invites citizens and policy makers to think again about expert claims of inevitability, and to retake the future as a space for ethical and democratic imagining.'

Sheila Jasanoff, Harvard Kennedy School, USA

'*Experiment Earth* is a book that is urgently needed. As human development becomes ever-more interwoven with the evolution of climate, Stilgoe asks a profound question: "What does it mean to take responsibility for global climate?" His answer is more than about climate and science, and more than about geoengineering technologies. It is about how we see ourselves as responsible human beings, exercising power, creativity and judgement in the world, whilst remaining accountable to each other.'

Mike Hulme, King's College London, UK

'To geoengineer or not to geoengineer the climate will be one of the defining science and environment policy questions of the next fifty years. In *Experiment Earth*, Jack Stilgoe provides an indispensable guide to the theories, politics and personalities which have shaped this emerging debate. With his unique perspective on the controversial SPICE project and the internal machinations of the Royal Society, Stilgoe digs beneath more superficial media coverage, to understand geoengineering as an experimental site for new approaches to the governance of technology and innovation. Entertaining, informative and insightful, this book should be read by all those who care about the future of science, democracy and the environment.'

James Wilsdon, University of Sussex, UK

'Climate engineering is a challenging subject to approach. One must walk the line between normalisation of what, to many, appears unthinkable and a manifesto for despair and inaction opposite the very real threat of climate change. This book struggles admirably with this tension: what it is like to work on an idea you hope never happens, and how could you ever control it? Stilgoe has been afforded access to the scientists working in this difficult arena, building trust and detailing our, and his, struggle to come to terms with the enormity of the problem. If you want to be inspired to wrestle with the intellectual challenges of how one might govern climate engineering technologies there may never be a better and more timely read than this.'

Matt Watson, University of Bristol, UK

Science in Society Series

Series Editor: Steve Rayner

Institute for Science, Innovation and Society, University of Oxford

The Earthscan Science in Society Series aims to publish new high-quality research, teaching, practical and policy-related books on topics that address the complex and vitally important interface between science and society.

Vaccine Anxieties
Global Science, child health and society
Melissa Leach and James Fairhead

Democratizing Technology
Risk, responsibility and the Regulation of Chemicals
Anne Chapman

Genomics and Society
Legal, ethical and social dimensions
Edited by George Gaskell and Martin W. Bauer

A Web of Prevention
Biological weapons, life sciences and the governance of research
Edited by Brian Rappert and Caitrìona McLeish

Nanotechnology
Risk, ethics and law
Edited by Geoffrey Hunt and Michael Mehta

Unnatural Selection
The challenges of engineering tomorrow's people
Edited by Peter Healey and Steve Rayner

Debating Climate Change
Pathways through argument to agreement
Elizabeth L. Malone

Business Planning for Turbulent Times
New methods for applying scenarios
Edited by Rafael Ramírez, John W. Selsky and Kees van der Heijden

Influenza and Public Health
Learning from past pandemics
Tamara Giles-Vernick, Susan Craddock and Jennifer Gunn

Animals as Biotechnology
Ethics, sustainability and critical animal studies
Richard Twine

Uncertainty in Policy Making
Values and evidence in complex decisions
Michael Heazle

The Limits to Scarcity
Contesting the politics of allocation
Lyla Mehta

Rationality and Ritual
Participation and exclusion in nuclear decision making, 2nd edn.
Brian Wynne

Integrating Science and Policy
Vulnerability and resilience in global environmental change
Edited by Roger E. Kasperson and Mimi Berberian

Dynamics of Disaster
Lessons on risk response and recovery
Rachel A. Dowty Beech and Barbara Allen

The Social Dynamics of Carbon Capture and Storage
Understanding CCS Representations, Governance and Innovation
Edited by Nils Markusson, Simon Shackley and Benjamin Evar

Science and Public Reason
Sheila Jasanoff

Marginalized Reproduction
Ethnicity, infertility and reproductive technologies
Edited by Lorraine Culley, Nicky Hudson and Floor van Rooij

Resolving Messy Policy Problems
Handling conflict in environmental, transport, health and ageing policy
Steven Ney

The Hartwell Approach to Climate Policy
Edited by Steve Rayner and Mark Caine

Reconstructing Sustainability Science
Knowledge and action for a sustainable future
Thaddeus R. Miller

Experiment Earth
Responsible innovation in geoengineering
Jack Stilgoe

Experiment Earth

Responsible innovation in geoengineering

Jack Stilgoe

Routledge
Taylor & Francis Group
LONDON AND NEW YORK

earthscan
from Routledge

First published 2015
by Routledge
2 Park Square, Milton Park, Abingdon, Oxon OX14 4RN

and by Routledge
711 Third Avenue, New York, NY 10017

Routledge is an imprint of the Taylor & Francis Group, an informa business

British Library Cataloguing-in-Publication Data
A catalogue record for this book is available from the British Library

Library of Congress Cataloging-in-Publication Data
A catalog record for this book has been requested

ISBN: 978-0-415-73237-6 (hbk)
ISBN: 978-1-315-84919-5 (ebk)

Typeset in Goudy
by Swales & Willis, Exeter, Devon, UK

MIX
Paper from responsible sources
FSC www.fsc.org FSC® C013604

Printed and bound by CPI Group (UK) Ltd, Croydon, CR0 4YY

Fear no more the heat o' the sun,
Nor the furious winter's rages;
Thou thy worldly task hast done,
Home art gone, and ta'en thy wages:
Golden lads and girls all must,
As chimney-sweepers, come to dust.

William Shakespeare, Cymbeline, Act 4, Scene 2

For Faith, Hope, Leo and Zachary

Contents

Figures

Table

Boxes

Acknowledgements

This book does not offer a final word on geoengineering. It draws on unfinished conversations about an issue that is still in formation. My view is that understanding and governing technologies at an early stage is vital because of, rather than despite, the uncertainties that characterise them. If we ever live in a geoengineered world, it won't look like the one currently being imagined, because prediction is always imperfect. Alternatively, geoengineering of the sort described in this book may come to be regarded as beyond the pale, as it has been in the past. Either way, parts of my analysis will quickly become out of date. My hope is that the broader lessons gleaned from close engagement with the world of geoengineering research remain relevant.

This book began as a research project funded by the Engineering and Physical Sciences Research Council (EPSRC) and the Economic and Social Research Council (ESRC). As well as funding the project, EPSRC deserves recognition for opening up their own processes to research and advice. ESRC and EPSRC both provided additional funding.

In addition to the two funders of this research, I owe huge thanks to the researchers in the SPICE project. As described in this book, their research involved their coming to terms with the emerging science of geoengineering, alongside its highly contested and surprising politics. At the same time they endured inconvenient conversations with me, accepted the substantial responsibilities that were thrust upon them and volunteered for more. Thanks above all to Matt Watson, who has embraced social science as a part of geoengineering research. His willingness to engage in an experimental collaboration when he had quite enough to worry about went well beyond his usual responsibilities. I would not expect any of the SPICE team to agree wholeheartedly with my analysis, but I would hope that they find something useful in it.

Thanks to all of the people who agreed to be interviewed as part of my research: Ken Caldeira, Mike Childs, Jolene Cook, Peter Davidson, Simon Driscoll, Pru Foster, Lesley Gray, Steven Hamburg, Clive Hamilton, Alex Hanafi, Jim Haywood, Richard Heap, Hugh Hunt, Peter Irvine, David Keith, Kirsty Kuo, Mark Lawrence, Tim Lenton, Andrew Lockley, Jane Long, Duncan McLaren, Scott Osprey, Andy Parker, Stuart Parkinson, Shobita Parthasarathy, Helena Paul, Steve Rayner, Alan Robock, Stephen Salter, David Santillo,

Hauke Schmidt, John Shepherd, Pablo Suarez, Jon Taylor, Matt Watson, Kathy Jo Wetter and David Winickoff. To preserve anonymity, quotations from interviews, which appear in italics in the text, are not identified by pseudonyms, codes or interview numbers. In addition to the insights provided by my interviewees, countless conversations with others involved in geoengineering research have informed the arguments here.

Thanks are also due to various audiences and interlocutors with whom I have explored half-baked versions of these ideas. This includes the students on my Responsible Science and Innovation and Governing Emerging Technologies courses at University College London.

Much of this research took place in collaboration with Richard Owen and Phil Macnaghten. Many of the ideas were developed in conversation with them. It has also been a privilege to work with Bron Szerszynski and Paul Oldham. Thanks, too, to all of the researchers on the ESRC Climate Geoengineering Governance project, who invited me to join their discussions and help my own research proceed in step with other social science on geoengineering. Thanks in particular are due to the leader of that project and editor of this book series, Steve Rayner, not just for supporting and guiding the production of this book, but also for helping to shape my thinking on so many of the issues I discuss. Steve was one of the first social scientists to realise how important and interesting geoengineering would turn out to be, and he has been instrumental in keeping the issue open enough for social scientists to engage constructively.

Earlier versions of some of the ideas in this book have been published in articles and reports. Chapter 2 uses and extends a report written for EPSRC: 'An Outline Framework for Responsible Innovation', by Stilgoe, Owen and Macnaghten (2012). These ideas were later published as 'Developing a Framework for Responsible Innovation', by Stilgoe, Owen and Macnaghten (2013). Chapter 5 also draws on this paper, in addition to reporting on research already published as 'Public Engagement with Biotechnologies Offers Lessons for the Governance of Geoengineering Research and Beyond', by Stilgoe, Watson and Kuo (2013).

Thanks to Jon Agar, Brian Balmer, Chris Brierley, Andy Parker, Rufus Stilgoe, Andy Stirling, Jon Turney, Matt Watson and James Wilsdon, all of whom read chapters and provided useful corrections and comments. Of course, none of these people deserve any responsibility for oversights, errors or shortcomings.

Acronyms and initialisms

CDR	carbon dioxide removal
CEO	chief executive officer
Defra	Department for Environment, Food and Rural Affairs
DICE	dynamic integrated climate–economy
ELSI	Ethical, Legal and Social Implications [US research program]
ENMOD	Environmental Modification Convention
E-PEACE	Eastern Pacific Emitted Aerosol Cloud Experiment
EPSRC	Engineering and Physical Sciences Research Council
ESRC	Economic and Social Research Council
EU	European Union
GeoMIP	Geoengineering Model Intercomparison Project
HSRC	Haida Salmon Restoration Corporation
IAGP	Integrated Assessment of Geoengineering Proposals
IPCC	Intergovernmental Panel on Climate Change
NanoCode	Code of Conduct for Responsible Nanosciences and Nanotechnologies Research
NERC	Natural Environment Research Council
NGO	non-governmental organisation
OIF	ocean iron fertilisation
PR	public relations
RAEng	Royal Academy of Engineering
SI	Système International d'Unités
SPICE	Stratospheric Particle Injection for Climate Engineering [UK project]
SPM	'Summary for Policymakers' [IPCC]
SRM	solar radiation management
SRMGI	Solar Radiation Management Governance Initiative
STFC	Science and Technology Facilities Council
STIR	Socio-Technical Integration Research [US project]
STS	science and technology studies
USSR	Union of Soviet Socialist Republics

1 Balloon debate

'When we all stand in that field in Norfolk, all of the engineers will be jumping up and down because they've succeeded in doing something amazing, building the tallest structure anywhere on Earth, and all of the natural scientists will be saying "Oh shit, we're a step closer to doing something bonkers".'

(A scientist working on the SPICE project)

A helium-filled polythene balloon floats three metres off the ground, tethered to a steel platform. The idea is to use this balloon to lift a kilometre-long hose into the sky. Once the balloon is up, some water – no more than it would take to fill a child's paddling pool – will be pumped up the hose and squirted out through a nozzle to form a fine mist. After a few test launches, the balloon will stay in the air for about five days, enough time for the engineers to observe how the apparatus withstands the wind: to see if the balloon dips, kites or spins and to see if the pipe twists, bends or wobbles.

There are two ways of looking at this experiment. From one perspective, it is a straightforward test of a combination of old, mundane technologies. The balloon is an 18-metre-long blimp, normally used at sporting events to hold TV cameras or advertising. It is not aiming that high. In the world of tethered balloons, the current altitude record is around five kilometres. The pump is from the sort of pressure washer that can be bought from a garden centre. The hose will be a longer version of the hydraulic hoses that carry fluids around a car. The small quantity of water means that it will probably evaporate before it hits the ground. The experiment will have no discernible effect on the environment.

The experiment has passed through two university ethics committees. The first responded that as the project did not involve animals or human subjects, it complied with ethical research standards. The second agreed, adding that the team's plans to engage members of the local community around the test site were welcome.

Such experiments are never risk-free. The engineers' own risk assessment points to a number of possible incidents. The balloon could deflate, perhaps because of a bird strike. High winds could drag the balloon back down to Earth. The winch could jam, leaving the balloon stuck in the sky. The tether could break free. (One of the engineers told me a story of a woman in California who had recently been pulled from her bicycle by a rogue rope from a hot-air balloon.)

It is important to bear these risks in mind, but such things are relatively well understood. Engineers have centuries of accumulated knowledge assessing and controlling risk. From a purely technical perspective, it is possible to conclude that nothing new is happening with this experiment. Few people outside the project are worried by the immediate risks. The non-governmental organisations (NGOs) and journalists who have taken an interest in this experiment are less concerned about the experiment going wrong than about it going right.

The second way of looking at this experiment is as '*the first field test of a geoengineering technology in the UK*', to use the researchers' own words. The experiment is part of a larger scientific project, known as SPICE. The playful acronym hides a serious motive – Stratospheric Particle Injection for Climate Engineering. One of the aims of this research is to work out whether it is possible to put particles into the stratosphere to reduce the amount of sunlight that reaches the Earth's surface. On the SPICE project's website, there is a schematic of a much larger balloon attached to a hose more than 20 kilometres long, spraying out a reflective aerosol that has yet to be determined but is likely to be less benign than water. Such a contraption is unachievable using present materials, but the design could be seen as a statement of intent.

The accepted definition of 'geoengineering' (or 'climate engineering') is the 'deliberate and large-scale intervention in the Earth's climatic system with the aim of reducing global warming' (Royal Society 2009, p. ix), through either sucking carbon dioxide from the air or reflecting sunlight back into space. Less than a decade ago, this big idea was given short shrift by both policymakers and scientists. The last five years have seen a dramatic increase in scientific interest. In September 2013, geoengineering was pushed closer to the mainstream of climate policy with a mention in the 'Summary for Policymakers' (SPM) of the fifth report of the Intergovernmental Panel on Climate Change (IPCC 2013).

The SPICE team are among a small but growing number of scientists taking the idea of geoengineering seriously. This is not to say that the SPICE scientists are trying to hasten a geoengineered future. They have, in the main, entered this new field with ambivalence and trepidation. The idea of geoengineering seems to cross Rubicons and break taboos. Some of the scientists are concerned that manipulating a system as chaotic and poorly understood as the global climate is likely to be disastrous. They point to early results from computer models that suggest dramatic effects on local weather patterns if global sunlight is reduced. Others point to the political risks of taking seriously a technological fix that destabilises the fragile political consensus on tackling climate change by cutting greenhouse gas emissions. Alan Robock, a climatologist, has produced an influential summary of 'reasons why geoengineering may be a bad idea' (Robock 2008). These concerns do not apply just to the use of any eventual technology. Given the potential downsides of this imagined technology, most scientists are at pains to emphasise that they would have no wish to deploy such a thing if it were developed. It is hard to find a geoengineering researcher who is in favour of doing geoengineering. But Robock and other scientists recognise that research on geoengineering may be a step onto a 'slippery slope', making technological development and deployment more likely (see also Jamieson 1996).

There are other reasons to be concerned about geoengineering that cannot be assessed by science but are no less important. If geoengineering of the type imagined by SPICE were to happen, it would represent a project of extraordinary hubris. It would concentrate power in the hands of very few people and claim mastery over a part of everyday life that we have until now been happy to admit is in some way out of our control. Even in our secular age, courts and insurance companies refer to extreme weather as an 'act of God'. An engineered climate would mean someone taking responsibility for such things. It is therefore reasonable to ask if this is the sort of world in which we would want to live. Many would legitimately respond that regardless of what the science tells us about risks and benefits, they would rather not head in that direction. It is in this sense that high-profile commentators express repugnance at geoengineering. The broadcaster David Attenborough has called the idea 'fascist',[1] an accessible if overstated recognition of what I and others have described as the anti-democratic political constitution of geoengineering proposals (Szerszynski *et al.* 2013).

Geoengineering is an emerging technology. We do not know precisely what a successful geoengineering device or technique will look like or how it will work. For now, geoengineering brings together a set of diverse proposals and suggestions. These range from the fantastic (sunshades in space between the sun and the Earth) to the well established (growing more trees or burying carbon dioxide underground). A couple of proposed geoengineering techniques have become the subjects of serious research. In addition to considering stratospheric particles, scientists have begun to experiment with ocean iron fertilisation. This involves the seeding of oceans with iron particles to encourage the growth of algae that would absorb carbon dioxide from the atmosphere and take it to the sea floor.

The experiment with the balloon is not attempting to do stratospheric particle injection, nor is it attempting to do climate engineering. But it is in some respects a 'climate experiment', as one journalist dubbed it.[2] A small group of campaigning NGOs issued a press release with the headline 'Say No to the Trojan Hose!' (ETC Group 2011). They wrote to the heads of the research councils and to government ministers, calling for the cancellation of an experiment that they saw as part of a rush to develop geoengineering.[3] Other geoengineering researchers around the world also criticised the haste with which the experiment seemed to be proceeding.

Both views of this open-air experiment are, in a strict sense, correct. But they reflect very different ways of understanding science in society. The first sees science in splendid isolation. The second sees scientific research entangled in the multiple lines of debate that characterise the geoengineering issue. The experiment was consciously public. It was announced at a national science festival with press releases and PR support from the universities involved. It revealed some of the assumptions and interests of geoengineering research to a wider audience for the first time. It therefore allowed for public scrutiny. The experiment, and the controversy it generated, provided a valuable opportunity for sociological research but also for what Arie Rip calls 'informal technology assessment' by those outside the scientific community (Rip 1986).

Our interest in scientific experiments need not be limited to those that take place outside or involve outsiders. Geoengineering of the sort under investigation by SPICE began as a set of thought experiments exploring the possibility of replicating the 'natural experiment' of a volcanic eruption. These ideas are now being tested using experiments run on computer models of the climate. We should take an interest in scientific research whatever its form, particularly when it is tied to such a problematic technological vision. The SPICE project is about much more than a balloon. The questions raised by *in vivo* or *in situ* experimentation can be reflected back on experiments taking place *in vitro* or *in silico*.

Conventionally, we regard thought experiments as constrained only by the scientific imagination. But, as I describe in later chapters, there are limits, norms and taboos that govern what scientists consider important, desirable or even thinkable. The future of the planet may be written in the experiments that take place inside laboratories, as much as outside. The direction of geoengineering research is a function of conversations that happen in public as well as those that involve just scientists. As geoengineering researchers start to take seriously the possibility of engineering the climate, which may profoundly recast humanity's relationship with the planet, we should look closely at dynamics of research, responsibility and governance.

This book is a sociology of geoengineering research. It draws on more than three years of interviews and interactions with the SPICE project and the wider geoengineering research community. It is about the tangle of issues in which geoengineering researchers find themselves. The book considers the various issues raised by geoengineering, focussing in particular on stratospheric particle injection, one of the subset of geoengineering proposals known as solar radiation management (SRM). It looks at how institutions and individuals have begun to make sense of solar geoengineering as it moves from the arena of science fiction into the arena of scientific research.

The book fits into the tradition of science and technology studies (STS), which is concerned with the social and political dimensions of science and engineering with a view to revealing the possibility of alternative directions. I am interested in the public nature of contentious science, its connections with emerging technologies and the negotiation of scientific responsibility. The publicness of geoengineering should make us pay attention not just to what is being done in the name of science, but also to how ideas of politics, ethics and 'the public' are being imagined. We should question the way that geoengineering is being framed as its complexities are made tractable through research and experimentation (cf. Bonneuil *et al.* 2008).

Governance beyond risk and ethics

The conclusions of recent STS studies of emerging technologies suggest the need for a rethink of the governance of science and innovation. We conventionally talk about the downsides of technology in terms of the risks or ethical dilemmas they create. The SPICE balloon debate was not really about either risk or ethics. STS research has revealed how this focus on the downstream impacts of technology can hide more fundamental upstream questions about the direction of innovation (Rayner and Cantor 1987; Wynne 2002).

With geoengineering, there is already plenty being said about risks. Scientists argue that observations of massive volcanic eruptions such as Mount Pinatubo in 1991 reveal both the cooling effect of particles in the stratosphere and the risks, particularly in the form of disruption to weather systems, when such an event happens.

Some see these risks as mountainous, even insurmountable. Others are more confident. For David Keith, currently the world's most prominent geoengineering researcher, volcanic eruptions 'give confidence that there is a strong empirical basis on which to assess these risks, and it is a reason to expect that the risks will be comparatively small' (Keith 2013, p. 12).

My aim in this book is to draw attention away from risk assessment towards uncertainties: the things we don't know, that we can't calculate and that may remain incalculable. Keith states that when it comes to the risks of climate change, 'we can't estimate the uncertainty very well: we don't know what we don't know' (Keith 2013, p. 31). The same applies to geoengineering. Keith admits that 'the largest concern is not the risks we know but rather a sensible fear of the unknown-unknowns that may surprise us' (Keith 2013, p. 70). For all this uncertainty, however, he is confident that science and engineering can find their way to a technology with 'negligible direct side effects' (Keith 2013, p. 110). Geoengineering researchers have begun taming some of these uncertainties and turning them into a research agenda. The assumption is that, as one paper claims, 'many uncertainties could be reduced through a systematic program of theory and modeling' (MacMynowski *et al.* 2011, pp. 5044–5045). STS research has demonstrated that in many areas, research creates more questions than answers, expanding our uncertainty (Nelkin 1979; Ravetz 1986). Uncertainty is just as important a part of science as knowledge is (Stocking 1998), and yet it is often hidden from public view. We can imagine that given the social and political complexities of geoengineering, the range of uncertainties is likely to be ever-expanding. Scientists should not pretend to completely know the risks and ethical challenges we face.

Science in society; science and society

This book is about the place of science, technology and innovation in the world. It is about 'science in society', but the conventional separation between 'science' and 'society' is one of many dichotomies challenged by my approach. Books like this are often categorised as 'science and society', as though these are worlds apart, or as 'science in society', as though science is a separate enclave. Despite the efforts of social scientists, the debate about science and innovation in society has struggled to avoid the implication that the science is somehow immutable, detached and exogenous. The logic follows that it is incumbent upon society and politics to understand, catch up and, if necessary, regulate.

The idea of geoengineering cannot be straightforwardly separated into scientific and social parts. The nascent debate about geoengineering shares some features with previous emerging technologies, including biotechnology and nanotechnology, which are driven by 'sociotechnical imaginaries' (Jasanoff and Kim 2009) – visions of desirable futures that blend social and scientific ambitions and carry narratives of both promise and threat. Their imagined potential demands government investment but also governance. It has become a commonplace of emerging technology discussions to identify a 'governance gap'.

David Keith prefaces his recent book, *A Case for Climate Engineering*, by stating, as if it were incontrovertible, the following:

> It is possible to cool the planet by injecting reflective particles of sulfuric acid into the upper atmosphere where they would scatter a tiny fraction of incoming sunlight back to space, creating a thin sunshade for the ground beneath. To say that it's 'possible' understates the case: it is cheap and technically easy.
>
> (Keith 2013, p. ix)

This argument, reminiscent of claims made at the dawn of nuclear power that people would be able to 'enjoy in their homes electrical energy too cheap to meter',[4] has been allowed to underpin assessments of the promises and perils of geoengineering. Keith goes on to evaluate the risks and benefits of the technology as though it were ready to be pulled off a shelf, with risk assessments all in order. He calls solar geoengineering a 'cheap tool that could green the world' (p. x) but argues for awareness of 'benefits and risks that are distributed at regional to global scales' (p. xx).

Claims about the potency of geoengineering have led to concerns about a gap between science and regulation:

- 'I think the science is certainly far out ahead of the politics' (Jason Blackstock, quoted in O'Neill 2012).
- 'Right now, the politics of geoengineering are far ahead of the science' (Victor *et al.* 2013).

These two quotes from geoengineering commentators, while apparently in disagreement, are actually pointing to the same thing – a technology that is neutral and inevitable, if not already present. The first is a call to govern the technology; the second is a call to better understand the technology. Both are deterministic. Neither admits that the technology remains the figment of a particular technoscientific imagination. This book hopes to contribute to the discussion of the governance of geoengineering by questioning the presence of this imagined governance gap. The STS critique of technological determinism first demands scepticism about the nature of the technology that is under investigation (Wyatt 2008).

The SPICE project does not come with easy distinctions between facts and values. Such distinctions are often made in the geoengineering debate, but that does not mean we should take them for granted. Part of my argument in this book is that a constructive debate about geoengineering requires recognition that its science and its politics have emerged hand-in-hand and will continue to do so. If we are to make sense of, learn from and deal with imaginaries of geoengineering, we should question the lines that are drawn between science and society, between nature and humanity and between research and innovation.

From noun to verb

Despite its non-existence as a viable sociotechnical system, geoengineering is already discussed as if it were a noun, an artefact. There is, oddly, rather little engineering in the world of geoengineering. The technology is simply imagined. Geoengineering, even in the absence of any concerted technology development, is talked about as if it is geotechnology, assessable using conventional geoscience. Geoengineering, as a noun, is becoming important in the climate change debate.

The SPMs provided by the IPCC are the frontline of negotiations between climate science and climate policy. In 2007, when the IPCC produced its fourth assessment report, geoengineering appeared only in the SPM for Working Group III, whose job is to assess options for mitigating climate change.

> Geo-engineering options, such as ocean fertilization to remove CO_2 directly from the atmosphere, or blocking sunlight by bringing material into the upper atmosphere, remain largely speculative and unproven, and with the risk of unknown side-effects. Reliable cost estimates for these options have not been published.
>
> (IPCC 2007, p. 15)

By 2013, geoengineering merited a longer mention:

> Methods that aim to deliberately alter the climate system to counter climate change, termed geoengineering, have been proposed. Limited evidence precludes a comprehensive quantitative assessment of both Solar Radiation Management (SRM) and Carbon Dioxide Removal (CDR) and their impact on the climate system. CDR methods have biogeochemical and technological limitations to their potential on a global scale. There is insufficient knowledge to quantify how much CO_2 emissions could be partially offset by CDR on a century timescale. Modelling indicates that SRM methods, if realizable, have the potential to substantially offset a global temperature rise, but they would also modify the global water cycle, and would not reduce ocean acidification. If SRM were terminated for any reason, there is high

> confidence that global surface temperatures would rise very rapidly to values consistent with the greenhouse gas forcing. CDR and SRM methods carry side effects and long-term consequences on a global scale.
>
> (IPCC 2013, p. 29)

This paragraph is the final one of the summary, giving the unfortunate impression of a twist in the climate change tale. A lazy policymaker might see it as an invitation to explore easy technological fixes rather than hard international negotiations. It is clear, however, that the IPCC sees little to like about geoengineering. The language has become more certain than in the previous report. Technologies that were 'speculative and unproven' in 2007 are now discussed, with relatively little additional evidence, in terms of their potential and their side effects. Tellingly, however, this paragraph is in the SPM from Working Group I, which assesses the 'physical science basis' for climate change. Solar geoengineering is not mentioned in the Working Group III SPM. It seems to have been naturalised as part of the physics of climate change, rather than explored as an engineering or policy option.

Engaging with geoengineering research in action forces us to think about geoengineering not as a noun but as a verb. Viewed as a set of technologies, geoengineering resembles no more than a mixed bag of half-baked schemes. If we take literally the meaning of 'geoengineering' as a present participle, it becomes a project, a work-in-progress. This idea, the idea of exerting control over the atmosphere, demands different sorts of analysis and governance. We can see geoengineering as a new trajectory, reconfiguring social relationships and, ultimately, reorienting people's relationship to the weather. My argument is that rather than talking about the governance of geoengineering, we should bring these things together. In its ambitions for climatic control, geoengineering is itself a form of governance.

This view – of geoengineering as a work-in-progress – changes our view of responsibility. As Oliver Morton (2012) has argued, researchers looking at geoengineering 'tend to naturalise it: to treat it as a thing in the world to be examined'. This leads towards arguments for technological and scientific autonomy. As I will describe, the naturalising of geoengineering has contributed to the framing of expert assessments of the issue and research agendas that have followed, in which scientists have been able to avoid many of the most profound questions of responsibility that geoengineering would seem to present.

If geoengineering is framed alternatively as a technoscientific project, the responsibilities of scientists are mixed: in 'researching' something, they are also implicated in its development, even if their research points to more risks than benefits. (This point applies equally to social scientists and others, me included, attracted to the purported novelties of geoengineering.)

Science is not one thing, nor does it have just one place in society. It is conflicted and its roles are multiple. It deals in truth, but also in innovation, expertise, evidence and critique.[5] Different disciplines, particularly if we include engineering, have different dispositions and come up with very different accounts of the world. It is increasingly hard to justify the activiti technoscience with reference to an old-fashioned model I hope to challenge the dominant framing of geoengineeri governed – by instead developing a narrative of the 'co-prod science and politics of geoengineering.

Understanding emerging technologies requires the di boundaries between science and society. Typically with a division is quickly established between those who want to innovate who want to regulate, with the reach of the former always exceeding the grasp of the latter.[6] With geoengineering, such cracks are only just starting to show. Few people are actively promoting the technology. Most geoengineering researchers are openly ambivalent about the technology and appreciate that the relevant questions reach far beyond science. As I will describe in subsequent chapters, however, this does not mean that the technology is stillborn. Some of the more thoughtful geoengineering researchers recognise that by researching something they see as highly undesirable, they may be unwittingly nurturing its development.

From speculation to anticipation

This is a book about geoengineering, but it is unlike other books about geoengineering. It does not share either the excitement or the terror evident in some of the books that have followed in the wake of scientific attention (Goodell 2010; Kintisch 2010; Hamilton 2013). These have all drawn attention to an important set of issues, but they have adopted the dominant scientific narrative of power and novelty that accompanies geoengineering debates.

We do not have to accept the *faits accomplis* suggested by the subtitles of those books. I do not see an 'audacious quest' (Goodell 2010) to engineer the Earth's climate, nor do I believe that we are witnessing 'the dawn of the age of climate engineering' (Hamilton 2013). The response to Kintisch's (2010) question of whether geoengineering is science's 'best hope' or 'worst nightmare' is almost certainly 'neither'. The choice of 'environmental necessity or Pandora's box' (Launder and Thompson 2010) is a false one.

James Fleming has taken a different approach in his history of geoengineering. His narrative of continuity from earlier, mainly spurious attempts at weather modification gives cause for scepticism about the novelty of geoengineering. For Fleming, most geoengineering science is 'geo-scientific speculation' (Fleming 2010, p. 228), based on 'back-of-the-envelope calculations' (p. 233). But the speculation is not just the preserve of scientists. We have seen an array of philosophers, legal scholars, social scientists and others gather to discuss the various non-technical issues that might arise. I confess that I share some of the fascination, which accounts for my writing this book. But the book is in part a criticism of what Alfred Nordmann (2007) has called 'speculative ethics'. As I describe in Chapter 3, there has been a minor explosion of science, social science and humanities research wargaming scenarios of a geoengineered future. There are discussions of who would be most likely to unilaterally use the technology, who [illegible] win and who would lose, how agreement might be reached on an ideal planetary temperature, and how planetary temperatures would rebound in the event of a technological shutdown. Ethicists have rushed to describe the questions of justice and rights that would arise from such scenarios.

The critique of speculative ethics is that such thinking cements the speculation, bringing it closer to inevitability. In discussing 'what will happen if . . .' the 'if' is more likely to become a 'when'. Nordmann describes how, as 'the hypothetical gets displaced by a supposed actual, an imagined future overwhelms the present' (Nordmann 2007, p. 32). With geoengineering, technologies are often discussed as though they are real. Researchers are already talking about whether the technology will be 'applied' (Barrett 2008) or 'deployed' (Victor 2008), rather than whether it can or should be developed.

Anticipating problems with the 'termination effect', the threat of 'unilateral deployment' and the control of the 'thermostat' has sparked important early discussions about the non-scientific aspects of geoengineering and its research, but such discussions risk exacerbating a narrow view of governance. With geoengineering, as with other emerging technologies, we should be concerned with its uses, as well as its abuses. Technological catastrophes may have rapid, visible and wide-ranging effects, but in the long run these are less important than the slow reconfigurations brought about by emerging technologies.

My straightforward response to the books which ask whether geoengineering will be the planet's saviour or a new disaster is 'we have no idea'. Mike Hulme takes a more critical approach to geoengineering science. He makes a strong argument that stratospheric aerosol injection is ungovernable, 'an illusory solution to the wrong problem' (Hulme 2014, p. 130) and therefore deserving of prohibition. He argues that 'the socio-technical imaginary of the thermostat should be dispensed with' (p. 82). I share many of his concerns, but we should not presume that the technologies currently imagined, with all of their hastily constructed 'implications', will come to be realised.

The danger is that, in speculating, we leapfrog the discussions in the present about how geoengineering research should proceed. Geoengineering is what Joel Mokyr (1990, p. 291) would call a 'hopeful monstrosity'. There are no technologies to see or touch, and the vast majority of scientific research has taken place inside computer models. The sociology of geoengineering is necessarily a sociology of ideas, promises, imagined futures and research trajectories. Geoengineering therefore provides a case study in what has been called the 'sociology of expectations' (Borup *et al.* 2006; Selin 2008).

Geoengineering futures

With an emerging technology, we typically see that the claims are grandest when the technology is least developed (Borup *et al.* 2006). In science, it might be reasonable to expect expectations to be at least partly backed up by evidence. But instead we typically see that the more immature the technology, the fewer constraints there are on hype. Futures are framed and constructed with stories, metaphors and clichés. Brigitte Nerlich and Rusi Jaspal point to the various linguistic devices with which actors have begun to make sense of this imaginary technology. The metaphors have joined the litany of narratives that already exist around climate change. Geoengineering is variously a 'dimmer switch', a 'thermostat', 'a sunshade', a 'plan B', a 'tool in scientists' toolbox', a 'parachute' in case of a planetary 'emergency'. Recognising the likelihood of side effects, geoengineering researchers have described it as a 'the lesser of two evils', 'chemotherapy' or 'methadone' for an addicted planet, with planet being a metaphorical body, machine or patient, according to the particular cliché. For critics, geoengineering represents a 'short term fix', a 'runaway technology', a 'moral hazard', 'playing God' or 'playing with fire' (Nerlich and Jaspal 2012). Clive Hamilton (2013) describes it as an archetypical 'Promethean' technology. In most cases, whether from the more techno-optimistic or critical ends of the spectrum, hope is accompanied by warning; hype sits alongside doom.[7] The clear message is that the technology is uniquely and unprecedentedly potent.

As with any new technology, there are definitional wrangles and frequent arguments for name changes and division of research areas. Some, such as adoption of the term 'climate remediation', suggest new ideas about what is desirable or technically plausible. These frames and futures are not just public relations. As one of my interviewees told me, '*framing is everything*'. It determines what seems acceptable or possible, who has a right to speak and the distribution of power (Schon and Rein 1994). My research is inspired by the sociology of expectations, but I do not presume that the futures imagined for geoengineering are either fixed or coherent. As with other areas of science that are accompanied by grand promises, from genomics to nanotechnology (Hedgecoe and Martin 2003; Nordmann and Rip 2009), trajectories of innovation can be modulated by new research, new controversies or new political arrangements.

If geoengineering is indeed 'a bad idea whose time has come' (Kintisch 2010, p. 13), we should ask why and how the promise of this idea has stabilised when a host of other grand technological schemes have been ridiculed, become relics of the Cold War or remained in the realm of science fiction. In the few years since geoengineering was rehabilitated as a credible topic of scientific research (see Chapter 3), geoengineering researchers have become increasingly self-confident. Doubts, uncertainties and ambivalences are being tamed. Ethical and political quandaries are being turned into empirical questions. Extraordinary proposals are being domesticated with ordinary science. The ease and cheapness of geoengineering is often taken for granted in geoengineering research. Geoengineering is often talked about as though it is an inevitable part of humanity's future relationship with the planet, and sometimes talked about as though it is already possible.

There are reasons why scientists such as David Keith pull a geoengineered future so close. Geoengineering is their object of study. Thankfully, it is neither as near nor as inevitable as Keith would have us believe. The sociotechnical system being imagined is highly uncertain, but we can expect the 'socio' part of it to be pretty important as it has proven to be with nuclear power, further compounding our uncertainties.

Geoengineering futures rest on assumptions about what is easy and what is hard, what is intractable and what is mobile. An important paper by Paul Crutzen (2006), discussed in Chapter 3, cemented the idea that the technology to cool the planet could be an easy solution to what has proven to be the hard if not impossible task of cutting greenhouse gas emissions. As I discuss in Chapter 2, geoengineering follows in a tradition of technological fixes that offer seductive alternatives to the difficult and messy business of policy, or what ardent technological fixer Alvin Weinberg (1966) called 'social engineering'.

The scale of ambition means that conversations about geoengineering can rapidly expand to encompass the future of the planet, the future of our species and humanity's relationship with Nature. The SPICE project brings such discussions back down to Earth. It prompts discussions of imagined, speculative and distant futures, but it demands attention to the immediate future too. The SPICE testbed experiment would have been one of the first experiments to test a geoengineering hypothesis outside a laboratory. The project attracted controversy for this reason, but it also created the possibility of unsettling assumptions that had come to dominate geoengineering futures. What if stratospheric geoengineering were more complicated, more expensive and more problematic than assumed?

The SPICE project has lessons for debates about geoengineering and debates about the governance of emerging technologies. But it is not just an interesting case study of scientific research. The SPICE balloon is also a symbol of the ambitions and flaws of contemporary science policy. As rich countries seek to secure their future as 'knowledge economies', science and scientists are under increasing pressure to contribute to economic growth. There is, as yet, no obvious capitalist aspect to solar geoengineering that is equivalent to the 'biocapital' (Sunder Rajan 2006) that now infuses the life sciences. Nevertheless, scientists still find themselves working under a regime of 'technoscientific promises' (Felt and Wynne 2007, p. 24) where, as Arie Rip puts it, 'being first is more important than going in the right direction' (Rip 2009). As I will describe in Chapter 5, the manner in which SPICE was funded displays some of this carelessness. But geoengineering offers an opportunity for an alternative view of the governance of innovation as 'collective experimentation' (Felt and Wynne 2007).

Collective experiments

This book turns on the idea of experimentation. Experiments are conventionally understood to be a scientific activity. But we have seen the term 'experiment' seep through the boundaries of science. It has become common to talk about the experimental nature of technologies that were once thought to be predictable and controllable (Krohn and Weyer 1994). And it is increasingly common to hear policy innovations described as 'experiments in governance' or 'experimental government'.

Experiments are normally part of the private life of science. The public image of science is about evidence, authority and expertise, not uncertainty and surprise, and when 'experiments' take place in public, they are typically displays of certainty (Shapin and Schaffer 1985; Collins 1988). If technologies are imagined as just things, society's questions are pushed downstream. With geoengineering, we see a clear need for democratic discussions to take place upstream (Wynne 2002; Wilsdon and Willis 2004), before we know what technologies will look like, what they will do and what they will mean for humanity. In this sense, geoengineering makes clear a need and an opportunity to democratise experimentation. SPICE provides an example of this happening in a semi-controlled fashion, with the gradual realisation that the outdoor experiment was about more than science.

The SPICE experiment was an early attempt to take geoengineering research out of the laboratory and into the field, from the domain of science to that of technology. The reaction to it took the scientists involved by surprise. Though originally intended as a technical test, it became a social experiment. Geoengineering is experimental in other ways, too. The planetary scale of ambition means that, as with nuclear power and other technologies, if a solar geoengineering technology is used, it will initiate a perpetual experiment with the planet. The technology can't be adequately tested until it is used – at scale and for a long time. The history of technology suggests that claims to be able to predict and control its effects are often overblown. But once society is locked into such experiments it becomes hard to withdraw from them. What begins as an experiment quickly becomes the everyday.

If we buy some of the arguments being put forward for the power of, for example, a future stratospheric aerosol technology, its potential for disruption puts it alongside nuclear weapons. But unlike the bomb, it won't just be created in secret and unleashed onto the world (unless we also buy the more far-fetched scenarios involving eco-terrorists or rogue states operating unilaterally). A geoengineered world, if we can imagine such a thing, would require a vast sociotechnical system of machinery, manpower, infrastructure, rules, laws and institutions. Innovation and experimentation will need to happen in public, with the public.

If geoengineering is to do what is expected of it, it will need to be tried, tested and scaled up. It will need to be experimented upon in the environment, *with* the environment, and the signal of its impact will need to be painstakingly extracted from the noise of a chaotic global climate. These tasks will be technically and politically difficult, and each will be fiercely contested. People will disagree about the shape, size and desirability of the experiments. And when they disagree, there will be further disagreements about who has given their consent to such experimentation. They will disagree about how to interpret the results. And once the technology is deemed – by whom, who knows? – worthy of deployment, the experiment will continue. The technology can only be tested through use, and the test will never provide uncontested certainty (MacKenzie 1990). A central insight from STS is that technologies and knowledge are never complete. Discussions about science and innovation can be closed, but their closure is done socially rather than technically. Such insights, and their occasional overuse, have attracted the accusation that STS is merely interested in deconstructing everything that we think might be solid. With geoengineering research, one does not have to look hard to see the bare bones of science. It is quickly apparent that there is a large range of views that are all in some way 'scientific', and there are plenty of scientists who admit the limits of science in understanding and charting a way forward. If we accept that innovation is somehow 'society in the making' (Callon 1987) and if we take research as an important part of 'innovation in the making', we should surely pay attention to the practice of scientific research.

Arguments about geoengineering are inextricably bound up with those about climate change, which has its own heavy political baggage. As I describe in Chapter 3, part of the commonplace rhetoric of anthropogenic climate change is that it represents an unprecedented 'experiment' with the Earth's climate. The implication is that this experiment has been an unethical one, but talk of this unplanned and uncontrolled experiment has made it easy for some technological optimists to suggest that what is needed is a controlled geoengineering experiment.

This is a book about what good experimentation might look like. If we consider the experimental system of geoengineering to include more than just scientists, their laboratories and their apparatus, how might we democratise experimentation? If we regard experimentation as a collective enterprise, something that is done *with* society rather than just *for* society or even *to* society, what are the responsibilities of scientists and the institutions that govern them?

About this book

This book is not an outsider's view of geoengineering. I do not pretend towards a scholarly detachment. I made the point above that all researchers interested in geoengineering may be in some way implicated in its future trajectory. I have sought to study geoengineering while being in some way part of it. The interactions that constitute the research behind this book are therefore themselves experimental, in the sense developed by Rabinow and Bennett (2012). The effect is that some of the reported conversations in this book are not the product of neutral observation but are snapshots of views at different points in a process of ongoing engagement, research and reflection.

The next chapter takes a step back from geoengineering to present a framework for considering whether and how science and innovation can take better care of the futures that they help bring about. I discuss the politics of technologies and technological fixes and advance notions of responsible innovation and collective experimentation.

Chapter 3 looks at the recent rise of geoengineering research, asking how a previously unthinkable area of science became 'thinkable'. I challenge the dominant history of geoengineering that has been adopted by geoengineering researchers, a story of disconnect from coventional climate science. The chapter draws threads together from the history of environmental science, the entangling of science and politics within the debate about climate change, and the mixed motivations for understanding, prediction and control within climate science.

Chapter 4 takes the Royal Society (2009) report on geoengineering as a case study in expert advice and technology assessment. I describe what was happening backstage as the Society wrestled with an issue that took the institution out of its scientific comfort zone. The Society's assessment was instrumental in the further construction of geoengineering. While the report, the Society staff and the working group were admirably open-minded in their approach, the issue became scientised in some important ways through their endorsement.

Chapter 5 looks in detail at the SPICE project, starting with the proposed outdoor experiment that initially attracted scrutiny. Asking what lessons about governance can be learnt from such experiences, I conclude that regulation of experimentation with clear lines around risk and ethical concern is unlikely to attract public credibility. We should instead seek to engage with the purposes of experimentation as part of a collective exploration of responsibility.

Chapter 6 looks at models in climate science and geoengineering research. I discuss the use of computer 'experiments' in climate science and ask what happens when the motivations for these experiments start to twist towards geoengineering. I look at the practice of climate science and the foibles of models that become visible up close.

Chapter 7 looks at dynamics of interdisciplinarity within and around the SPICE project. I ask how science fares when unfamiliar research cultures clash in new and contested areas. I consider the 'engineeringness' of geoengineering and argue that the disruption that comes from these forms of collaboration can be healthy as a precursor for taking greater responsibility.

In the book's concluding chapter, I consider the potential for democratising the collective experiment of geoengineering and offer suggestions for improved governance and careful research.

Taking responsibility

The story about the balloon experiment needs an ending. After lengthy discussions within the team and with the funders and others, the SPICE team decided not to launch the balloon. A patent application that included two of the SPICE researchers was unearthed, fuelling disagreements within the team about the merits of the experiment. Following an earlier postponement, and in recognition of the complexity of issues that had been surfaced by the proposal, the researchers called off the test.

The experimental gallimaufry was left unbuilt. The balloon that would have carried the hose was redeployed to some other task. The order for the hose was cancelled. The engineers on the SPICE project turned their attention to other ways of investigating the potential for giant stratospheric balloons, and the rest of the SPICE team continued with their research, albeit slowed by the administrative burdens of dealing with the fallout from the proposed experiment. The experiment had become a topic of conversation at all levels of British science policy, from the government chief scientific adviser downwards. The decision to cancel the testbed attracted wide news coverage, particularly in scientific publications such as *Nature*,[8] and initiated a period of soul-searching among scientists regarding what was at stake in geoengineering research.

The proposed SPICE experiment, the controversy it generated and the scientists' decision not to carry it through, all of which the SPICE scientists subsequently labelled 'the SPICE experience', prompt questions about responsibility that have become a central theme of this book. At the most abstract level, geoengineering raises the question of whether we are ready to take responsibility for the climate and therefore for the weather. But there are more immediate questions of responsible research and innovation that have been ignored for too long in cultures and institutions of science. At the time of writing, some solar geoengineering researchers are concocting and starting to propose a new set of outdoor geoengineering experiments. These researchers claim to have understood the lessons from SPICE, but it is clear that some lessons have been easier to hear than others. The lessons for responsible experimentation from this case are more profound than is immediately apparent.

As part of an extension of the typical risk-and-ethics model of governance described above, responsibility is often understood in the legalistic, retrospective sense of blame. As I explain in the next chapter, such a view reflects an impoverished view of governance in science. Science is, especially at its frontiers, largely self-governing. With emerging technologies, scientists are setting rules and norms as much as following them. We should pay attention to vested interests and any conflicts that may arise, but explaining the politics of geoengineering research does not require the construction of a conspiracy. Jane Long and Dane Scott have identified four vested interests that might contribute to shaping the future of geoengineering – fortune, fear, fame and fanaticism (Long and Scott 2013). To these we might add 'fascination', the everyday curiosity that drives scientists and other researchers to explore and in doing so construct the technoscience of geoengineering.[9] Many geoengineering researchers ventured into the area precisely because of a concern that others were seeking to geoengineer the planet. They now worry that their research may in some way be hastening a future they don't want to see. Some have forced themselves to keep an open mind about the desirability of doing geoengineering. Others display a more shameless enthusiasm. David Keith admits that in the geoengineering community,

> we're hiding a genuine and I think not-wrong joy in the fact that we understand something about the world that potentially gives us the ability to do these things. That understanding that nature gives us power to do great harm as well as, potentially, power to do good. But the understanding is a triumph of human ingenuity and I think it deserves some celebration although people are afraid to do that.
>
> (David Keith, quoted in NPR/TED Staff 2013)

Most geoengineering researchers make more pragmatic arguments for research. The basis for many of these arguments is inevitability: either the planet will need geoengineering, or the technology will be used unilaterally. In either case, as Granger Morgan puts it, 'If we haven't done the research . . . the international community has to fall back on a moral argument, as opposed to a science-based argument' (quoted in Inman 2010). The binary choice placed before society is between knowledge and ignorance, between an accelerator and a brake. My argument challenges such a simplification and asks instead what sorts of directions and qualities we might look for in responsible research.

Notes

1 Sir David Attenborough, speaking on the BBC's *The Andrew Marr Show*, 11 December 2011.
2 A story by John Vidal, 'Giant pipe and balloon to pump water into the sky in climate experiment', on the *Guardian* website, 31 August 2011, 16:00 BST, although the print edition had a different headline ('Want to mimic a volcano to combat global warming? Launch a Wembley-size balloon').
3 SPICE opposition letter, 26 September 2011. Available online at http://www.handsoff-motherearth.org/hose-experiment/spice-opposition-letter/ (accessed 29 July 2014).
4 Lewis Strauss, chairman of the United States Atomic Energy Commission, in a 1954 speech to the National Association of Science Writers.
5 Sheila Jasanoff (2014) refers to the tension between truth and 'gain' in science.
6 Approaches such as Constructive Technology Assessment try to bring promoters and controllers together into a common project (Schot and Rip 1997).
7 This dynamic has also been observed with nanotechnology and synthetic biology. See, for example, Ginsberg *et al.* (2014).
8 An incomplete list includes the following: Daniel Cressey (2012), 'Cancelled project spurs debate over geoengineering patents'; Geoff Brumfiel (2012), 'Good science, bad science'; *The Economist* (2012), 'Implicit promises: a geoengineering experiment has come unstuck. But there will be more'; Clive Cookson (2012), 'Scientists call off geoengineering trial'; and Mark Brown (2012), 'First test of floating volcano geoengineering project cancelled'.
9 David Santillo from Greenpeace is to be credited with this fifth 'F'.

References

Barrett, S. (2008). The incredible economics of geoengineering. *Environmental & Resource Economics*, 39(1), 45–54.

Bonneuil, C., Joly, P-B., and Marris, C. (2008). Disentrenching experiment: the construction of GM-crop field trials as a social problem. *Science, Technology, & Human Values*, 33(2), 201–229.

Borup, M., Brown, N., Konrad, K., and van Lente, H. (2006). The sociology of expectations in science and technology. *Technology Analysis & Strategic Management*, 18(3/4), 285–298.

Brown, M. (2012). First test of floating volcano geoengineering project cancelled. *Wired*, 16 May.

Brumfiel, G. (2012). Good science, bad science. *Nature*, 484(7395), 432–434.

Callon, M. (1987). Society in the making: the study of technology as a tool for sociological analysis. In W. E. Bijker, T. P. Hughes and T. Pinch (eds), *The social construction of technological systems: new directions in the sociology and history of technology* (pp. 83–103). London: MIT Press.

Collins, H. M. (1988). Public experiments and displays of virtuosity: the core-set revisited. *Social Studies of Science*, 18(4), 725–748.

Cookson, C. (2012). Scientists call off geoengineering trial. *Financial Times*, 16 May 2012.
Cressey, D. (2012). Cancelled project spurs debate over geoengineering patents. *Nature*, 485(7399), 429.
Crutzen, P. J. (2006). Albedo enhancement by stratospheric sulfur injections: a contribution to resolve a policy dilemma? *Climatic Change*, 77(3/4), 211–220.
Economist, The (2012). Implicit promises: a geoengineering experiment has come unstuck. But there will be more. *The Economist*, 19 May.
ETC Group (2011). *Say no to the 'Trojan hose': no spice in our skies, say environmental justice groups*. News Release, 27 September. Ottawa: ETC Group. Available online at http://www.handsoffmotherearth.org/wp-content/uploads/2011/09/NR-SPICE-270911.pdf (accessed 19 December 2013).
Felt, U., and Wynne, B. (2007). *Taking European knowledge society seriously*. Luxembourg: European Commission, DG for Research.
Fleming, J. R. (2010). *Fixing the sky: the checkered history of weather and climate control*. New York: Columbia University Press.
Ginsberg, A., Calvert, J., Schyfter, P., Elfick, A., and Endy, D. (2014). *Synthetic aesthetics: investigating synthetic biology's designs on Nature*. Cambridge, MA: MIT Press.
Goodell, J. (2010). *How to cool the planet: geoengineering and the audacious quest to fix Earth's climate*. Boston: Houghton Mifflin Harcourt.
Hamilton, C. (2013). *Earthmasters: the dawn of the age of climate engineering*. New Haven, CT: Yale University Press.
Hedgecoe, A., and Martin, P. (2003). The drugs don't work: expectations and the shaping of pharmacogenetics. *Social Studies of Science*, 33(3), 327–364.
Hulme, M. (2014). *Can science fix climate change? A case against climate engineering*. Cambridge: Polity Press.
Inman, M. (2010). Planning for plan B. *Nature Reports: Climate Change*, 4(Jan), 7–9.
IPCC (2007). Summary for policymakers. In B. Metz, O. R. Davidson, P. R. Bosch, R. Dave and L. A. Meyer (eds), *Climate change 2007: Mitigation of climate change*. Contribution of Working Group III to the Fourth Assessment Report of the Intergovernmental Panel on Climate Change. Cambridge: Cambridge University Press.
——— (2013). Summary for policymakers. In T. F. Stocker, D. Qin, G-K. Plattner, M. Tignor, S. K. Allen *et al*. (eds), *Climate change 2013: the physical science basis*. Contribution of Working Group I to the Fifth Assessment Report of the Intergovernmental Panel on Climate Change. Cambridge: Cambridge University Press.
Jamieson, D. (1996). Ethics and intentional climate change. *Climatic Change*, 33(3), 323–336.
Jasanoff, S. (ed.) (2004). *States of knowledge: the co-production of science and social order*. London: Routledge.
——— (2014). *Biotechnology and imaginaries of resistance*. Lecture, STEPS Centre Summer School, University of Sussex, Brighton, 18 May 2012. Available online at http://community.eldis.org/.5ae27939 (accessed 31 March 2014).
Jasanoff, S., and Kim, S-H. (2009). Containing the atom: sociotechnical imaginaries and nuclear power in the United States and South Korea. *Minerva*, 47(2), 119–146.
Keith, D. (2013). *A case for climate engineering*. Cambridge, MA: MIT Press.
Kintisch, E. (2010). *Hack the planet: science's best hope – or worst nightmare – for averting climate catastrophe*. Hoboken, NJ: Wiley.
Krohn, W., and Weyer, J. (1994). Society as a laboratory: the social risks of experimental research. *Science and Public Policy*, 21(3), 173–183.
Launder, B., and Thompson, J. M. T. (eds) (2010). *Geo-engineering climate change: environmental necessity or Pandora's box?* Cambridge: Cambridge University Press.

Long, J. C. S., and Scott, D. (2013). Vested interests and geoengineering research. *Issues in Science and Technology*, 29(3), 45–52.

MacKenzie, D. (1990). *Inventing accuracy: a historical sociology of nuclear missile guidance*. Cambridge, MA: MIT Press.

MacMynowski, D. G., Keith, D. W., Caldeira, K., and Shin, H. J. (2011). Can we test geoengineering? *Energy & Environmental Science*, 4(12), 5044–5052.

Mokyr, J. (1990). *The lever of riches: technological creativity and economic progress*. New York: Oxford University Press.

Morton, O. (2012). *On geoengineering*. TheEnergyCollective.com. Available online at http://theenergycollective.com/breakthroughinstitut/94851/geoengineering (accessed 24 April 2014).

Nelkin, D. (1979). *Controversy: the politics of technical decisions*. Beverly Hills: Sage.

Nerlich, B., and Jaspal, R. (2012). Metaphors we die by? Geoengineering, metaphors, and the argument from catastrophe. *Metaphor and Symbol*, 27(2), 131–147.

Nordmann, A. (2007). If and then: a critique of speculative nanoethics. *NanoEthics*, 1(1), 31–46.

Nordmann, A., and Rip, A. (2009). Mind the gap revisited. *Nature Nanotechnology*, 4(5), 273–274.

NPR/TED Staff (2013). *Can hacking the stratosphere solve climate change?* Available online at http://wap.npr.org/story/209191273 (accessed 20 November 2013).

O'Neill, M. 2012. Science fiction gets real: a special report. *Lateline*. Australian Broadcasting Corporation. Available online at http://www.abc.net.au/lateline/content/2012/s3639093.htm (accessed 20 November 2013).

Rabinow, P., and Bennett, G. (2012). *Designing human practices: an experiment with synthetic biology*. Chicago: University of Chicago Press.

Ravetz, J. (1986). Usable knowledge, usable ignorance: incomplete science with policy implications. In W. C. Clark and R. E. Munn, *Sustainable development of the biosphere* (pp. 415–432). New York: Cambridge University Press. Reprinted in 1987 in *Science Communication*, 9(1), 87–116.

Rayner, S., and Cantor, R. (1987). How fair is safe enough? The cultural approach to societal technology choice. *Risk Analysis*, 7(1), 3–9.

Rip, A. (1986). Controversies as informal technology assessment. *Science Communication*, 8(2), 349–371.

——— (2009). Futures of ELSA: Science & Society Series on Convergence Research. *EMBO Reports*, 10(7), 666–670.

Robock, A. (2008). 20 reasons why geoengineering may be a bad idea. *Bulletin of the Atomic Scientists*, 64(2), 14–18, 59.

Royal Society (2009). *Geoengineering the climate: science, governance and uncertainty*. London: Royal Society.

Schon, D., and Rein, M. (1994). *Frame reflection: toward the resolution of intractable policy controversies*. New York: Basic Books.

Schot, J., and Rip, A. (1997). The past and future of constructive technology assessment. *Technological Forecasting & Social Change*, 54(2/3), 251–268.

Selin, C. (2008). The sociology of the future: tracing stories of technology and time. *Sociology Compass*, 2(6), 1878–1895.

Shapin, S., and Schaffer, S. (1985). *Leviathan and the air-pump: Hobbes, Boyle, and the experimental life*. Princeton: Princeton University Press.

Stocking, S. H. (1998). On drawing attention to ignorance. *Science Communication*, 20(1), 165–178.

Sunder Rajan, K. (2006). *Biocapital: the constitution of postgenomic life*. Durham, NC: Duke University Press.

Szerszynski, B., Kearnes, M., Macnaghten, P., Owen, R., and Stilgoe, J. (2013). Why solar radiation management geoengineering and democracy won't mix. *Environment and Planning* A, 45(12), 2809–2816.

Victor, D. G. (2008). On the regulation of geoengineering. *Oxford Review of Economic Policy*, 24(2), 322–336.

Victor, D. G., Morgan, M. G., Apt, J., Steinbruner, J., and Ricke, K. (2013). The truth about geoengineering: science fiction and science fact. *Foreign Affairs*, 92(2), Postscript.

Weinberg, A. M. (1966). Can technology replace social engineering? *Bulletin of the Atomic Scientists*, 22(10), 4–8.

Wilsdon, J., and Willis, R. (2004). *See-through science: why public engagement needs to move upstream*. London: Demos.

Wyatt, S. (2008). Technological determinism is dead; long live technological determinism. In E. J. Hackett, O. Amsterdamska, M. E. Lynch and J. Wajcman (eds), *The handbook of science and technology studies*, 3rd edn (pp. 165–180). Cambridge, MA: MIT Press.

Wynne, B. (2002). Risk and environment as legitimatory discourses of technology: reflexivity inside out? *Current Sociology*, 50(3), 459–477.

2 Taking care of the future

On 12 February 1908, six cars gathered in Times Square, New York, surrounded by more than 250,000 spectators, for the start of the New York to Paris automobile race. Rather than being shipped across the Atlantic, these cars were going the long way round – west across the United States, transported on a ship to Japan and then Russia, before driving almost the entire width of Eurasia. Seven cars had withdrawn before the start, and three more pulled out during the race. Travelling much of the way without roads, the cars were forced to drive cross-country through half-frozen Siberian mud or along railways, with the drivers conducting running repairs along the route.

These cars were rudimentary but recognisable prototypes of their current equivalents. A German car had been constructed especially for the race, on the orders of Kaiser Wilhelm, but the US entry, a Thomas Flyer, was a regular production model. The Flyer won the race on 30 July, after the German car, which had finished ahead, was penalised for taking various shortcuts.[1] The race was a fanfare for a technology that would take over the world. Three months later, the first Model T Ford rolled off the production line that went on to define the modern factory.

In 1908, the motor car was an emerging technology. None of the race's drivers nor Henry Ford himself could have possibly predicted in 1908 the eventual ways in which our lives would be reshaped by and reorganised around cars. Early growth was explosive. In 1908, less than 0.2 per cent of the US population owned a car. By the 1920s, the proportion exceeded 10 per cent. In the 1950s, by which time a third of Americans owned cars, the geography of cities like Los Angeles visibly reflected the country's intimate dependence on the technology.[2] The car created immense opportunities. The average American in 1800 travelled 50 metres a day. The average distance is now 50 kilometres (Urry 2007). But our increasing dependence on the car has been accompanied by a growing realisation of its problems.

In 1963, by which time the UK had 7 million cars on its roads, a report commissioned from Sir Colin Buchanan by the Ministry of Transport reflected the systemic problems of car ownership and the deep ambivalence it induced:

> We are nourishing at immense cost a monster of great potential destructiveness. And yet we love him dearly. Regarded in its collective aspect as 'the traffic problem' the motor car is clearly a menace which can spoil our civilisation. But translated into terms of our own car, we regard it as one of our most treasured possessions or dearest ambitions, an immense convenience, an expander of the dimensions of life, an instrument of emancipation, a symbol of the modern age.
>
> (Buchanan 1963, p. 15)

This 'instrument of emancipation' brings a 'pandemic cataclysm', in J. G. Ballard's words,[3] of more than 1.2 million deaths per year.[4] In rich countries, cars and societies have evolved together such that they are largely interdependent. Although the centres of some cities that predate cars resist the technology's thrall, places like Phoenix, Arizona, owe their sprawling geography almost entirely to the car. We are 'locked in' to the sociotechnical system of automobility (Urry 1999), and many developing countries are following similar trajectories. Cars are not just things. They are part of a vast system. According to John Whitelegg, with our increasing car use, 'more money must be spent on roads, car-parking and all the associated infrastructure of dependency on motorised transport including the police and courts. Henry Ford would not have been impressed by the monster that he was instrumental in creating' (Whitelegg 1997, p. 18).

Perhaps the biggest problem is our inability to extricate ourselves. This system, which emerged alongside the innovation of the conventional internal-combustion-engined car, supports its existence and crowds out alternative options such as electric cars, public transport or cycling. This technology, once imagined as a servant of human needs, would seem to be taking over. It has become one of the 'things that we didn't know we couldn't do without' (Sudjic 2009, p. 60).

So, given the power of technological innovation to shape our lives, how might we anticipate, understand and govern the emergence of new technologies? In a landmark book, *The Social Control of Technology*, David Collingridge uses the car to illustrate what he calls the 'dilemma of control' – the implications of a technology are hard to predict in its early stages, but as we come to terms with them, the technology becomes more entrenched and therefore harder to control (Collingridge 1980). Collingridge considers an early attempt at automobile technology assessment, the Royal Commission on Motor Cars in 1906, two years before the New York to Paris race and the first Model T Ford. Though it did recommend a penalty for those found driving while drunk, the Royal Commission identified dust kicked up from untarred roads as the most serious problem. Given what we now know about cars, 'with hindsight we smile, but only with hindsight' (Collingridge 1980, p. 16).

Our ability to turn the lessons of hindsight towards improved foresight and governance is hampered by the dilemma of control that Collingridge (1980) describes. Viewed one way, we are destined to be both too early and too late in the governance of technology: too early to understand consequences and too late to steer technology away from certain trajectories. But decisions that are taken at the early stages of technological development, whether regarding cars or geoengineering, will have profound consequences downstream. Although we overestimate our ability to control technologies once they are fully formed, we underestimate our ability to shape science and innovation while they are still emerging.

Taking responsibility for the unpredictable nature of innovation is an enormous challenge, but it is a vital one. Bruno Latour (2011) advises us to 'love our monsters', not in the sense of being enamoured of our technological creations, but in the parental sense of caring about them and taking responsibility for their development. To do this, however, we need to recognise that such monsters are less monstrous than is sometimes assumed.

Prometheus and the golem

Although we cannot yet point to the artefacts of geoengineering as we could to the motorcar at the start of the twentieth century, we can nevertheless understand geoengineering as an emerging technology. As with other emerging areas of science and technology, we do not fully know what its technological products will look like, nor what their implications might be. And yet we must find some way to govern the progress of research and innovation, some way of proceeding responsibly.

Research in science and technology studies (STS) over more than four decades has demonstrated that research and innovation do not follow a simple internal logic. They are shaped by social as well as technical considerations. Technologies do not spring fully formed from their inventors' hands. They are socially shaped in their creation and in their subsequent use (Bijker and Law 1992). Technologies may do what innovators intend, but they will do other things besides. For Langdon Winner, the imperative to govern technologies democratically starts with the argument that they are a form of 'legislation', controlling our lives in various ways, only some of which we immediately appreciate (Winner 1977; also see M. B. Brown 2007).

We have learnt, too late, that technologies have unintended, potentially disastrous consequences. Some of these may have been unpredictable, but in most cases, including prominent ones such as asbestos and thalidomide, hazards were anticipated and detected early, but not acted upon (Harremoës *et al.* 2001). If technologies are to improve our lives, we might at least demand some form of Hippocratic Oath that they should first do no harm. But technologies, by creating new capabilities, always change social relationships, replace jobs and reshape lives. The distribution of risks and benefits from technology is uneven and often unpredictable. When we add the complex risks that come from a sociotechnical system like that surrounding the car, for which the car itself cannot be held responsible, it is tempting to conclude that technology is largely ungovernable.

In a world of pervasive technologies, of which democratic society appears to have a dwindling degree of understanding and control, the idea that technology is in some way autonomous seems to resonate. According to Jacques Ellul

(1962), the autonomy of technology has already enslaved us. The list of those who have made similar diagnoses and predictions includes not just philosophers and sociologists of technology, but also terrorists such as the Unabomber and, more recently, the founder of Sun Microsystems, Bill Joy. Joy argued, with reference to the Unabomber's manifesto and reflecting on recent developments in robotics, genetic engineering and nanotechnology, that 'we are being propelled into this new century with no plan, no control, no brakes' (Joy 2000).

Technology writer Kevin Kelly also claims that technology is autonomous, but he is optimistic. For Kelly, the task is to speed up the process of emancipatory innovation. His response to technological autonomy is to give up attempts at control, to prohibit prohibition, which only delays the inevitable (Kelly 2010). Bill Joy's view is less rosy, but he is similarly fatalistic about our ability to control science and technology. He argues that 'ideas can't be put back in a box . . . Once they are out, they are out' (Joy 2000). This is a Promethean narrative of scientific and technological autonomy. It resonates with another mythical narrative, of science as a golem, a 'lumbering fool who knows neither his own strength nor the extent of his clumsiness and ignorance' (Collins and Pinch 1998, p. 2). Such stories leave little room for questions of responsibility in innovation, beyond the metaphorical decision of whether to steal fire or create a monster. But they also reduce the question of technological control to one of prohibition. This has led to an unproductive discussion of the merits of precautionary regulation and its new libertarian counterpart, the so-called proactionary principle (see, for a discussion, Fuller 2012).

In the aggregate, science and technology may well share some characteristics of an autonomous life form (Winner 1983). But this view downplays the ways in which, at a smaller scale, people shape technologies (Bijker and Law 1992). The attempt to reconcile these things, and recover some hope for reasserting control of technology, has led to the theory of technological 'momentum' (Hughes 1993), in which technologies that begin as controllable become unstoppable as they are adopted. Bruno Latour puts forward the idea, following Akrich, that technologies are 'scripts'. For Latour, the making of technologies is about 'transcription' or 'inscription'. They 'prescribe' certain things from their users and 'proscribe' others. Sometimes the rules are clear, as with mundane technologies like the door-closer (Latour, writing as Jim Johnson, 1988; Latour 1992), the safety belt (Latour 1992) or the sleeping policeman (Latour and Venn 2002). Sometimes the instructions are more opaque, as with Latour and Venn's (2002) filing cabinet.[5] The inflexibility of technology leads Latour to conclude that 'no human is as relentlessly moral as a machine' (Latour, writing as Jim Johnson, 1988, p. 301). Others, most notably Langdon Winner, have argued that technologies contain implicit ethics and politics in what they allow and constrain. Peter-Paul Verbeek (2006, p. 361) argues that engineers therefore conduct 'ethics by other means'. But all these authors – Verbeek, Latour and Winner – recognise that the future of technologies is not written indelibly. Technologies are defined not just in their creation, but also by their use. Technologies can be hacked, adapted and repurposed. In many cases, we can choose not to subscribe to a technology or to the image of ourselves that it demands. We can change the scripts. Ronald Kline and Trevor Pinch (1996) describe how users of cars and radios in America reshaped the use and so the innovation of these technologies. Eric von Hippel (2005) and colleagues (Lüthje *et al.* 2005) explain how the mountain bike was a product of user-driven innovation in California in the 1970s and 80s. Winner concludes that we do not have to sleepwalk into a technological future:

> The key question is not how technology is constructed but how to come to terms with ways in which our technology-centered world might be reconstructed. Faced with a variety of social and environmental ills, there is growing recognition that what is needed is a process of redirecting our technological systems and projects in ways inspired by democratic and ecological principles.
>
> (Winner 1993, p. 376)

With such technologies, however, it would be naïve to suggest that users have all the power. As I said at the start of this chapter, the car locks us into a sociotechnical system that dramatically narrows our options. Other technologies can be seen as similarly authoritarian. We have little choice whether to subscribe to closed-circuit television, airport security, nuclear power or national defence, for example. In such cases, however, to emphasise the autonomy of these technologies is to miss the vast amount of effort and thought that goes into their creation. We might do well to remember, for example, that the atomic bomb required $2 billion and 200,000 people to turn a 'Promethean' idea into a workable technology (Miller and Edwards 2001).

If we are looking to govern technology while it is still emerging, when it is characterised by uncertainty, competing claims to novelty, promise and threat, and high political stakes, we should take seriously the concept of emergence, which holds that characteristics of the whole cannot be reduced to, nor predicted from, the sum of the parts. But, as Ilya Prigogine, the Nobel Prize-winning theorist of complexity, has argued (Prigogine 1997), this should give us cause for optimism. Just as emergence challenges the determinism of reductionist science, so it challenges the fatalism that can, for some, be the next step from ideas of technological autonomy.

Directions of innovation

In 1959, C. P. Snow argued in a forgotten sidenote to his famous Rede Lecture on the Two Cultures that the power of science and innovation was such that it would cure poverty. He made the following prediction: 'This disparity between rich and poor has been noticed . . . Whatever else survives to the year 2000, that won't' (see Stilgoe 2013). Rather than dwelling on the hazards of prediction, we should instead consider why it is that science and innovation appear to have, over the 50 years since Snow's lecture, failed to follow the direction he suggested. Richard Nelson (2011) rephrased the question: Why is it that our innovations have been able to put human beings on the surface of the moon but proven incapable of tackling the problems of urban poverty in the world's richest country?

The last decade has seen increasing policy attention given to big societal challenges.[6] The emergence of programmes of research based on 'grand challenges' reflects a desire to maximise the social and economic impact of science and technology. The challenges that are normally listed are familiar ones. The list normally includes sustainability of the global environment; the health of its population; the security of food, water and energy; and future sources of economic growth. For all the talk, however, there has been little progress in thinking about how science and innovation might be systematically turned in the direction of these problems.

We know that innovation is not a straight line in which new knowledge pushes new applications, which in turn lead to economic growth or social benefit. But this 'linear model' is continually rehearsed in the stories that science tells about its impact on the world (Felt and Wynne 2007). Even in areas of the economy where innovation seems at first glance to be more linear, such as pharmaceutical research, this model is falling apart. It is more realistic, and more helpful, to think of innovation not as a line but as a network or, according to some, an ecosystem. The innovation system is, according to one definition, 'the network of institutions in the public and private sectors whose activities and interactions initiate, import, modify and diffuse new technologies' (Freeman 1987, p. 1). In hindsight, the trajectory of certain innovations may look linear, because we can't see the dead ends, the technological failures, the negative results or the roads not taken. But the particular technologies that we currently have – whether internal-combustion engines, nuclear power stations, space shuttles, Microsoft Word or the QWERTY keyboard – are the product of choices, priorities, assumptions and desires through the innovation system. Alternative trajectories, in which people drive electric cars using electricity from cheap solar cells or in which more diseases of poverty have been eradicated, are imaginable, but they may be mutually exclusive. Innovation, which is often discussed by policymakers as just something that can either be fast or slow, also has a direction (Stirling 2008).

Technological fixes – good, bad and ugly

Given the problems facing society, many of which are 'wickedly' intractable (Rayner 2012), technological fixes remain seductive, even though we have learnt how disappointing they can often be. The phrase comes from Alvin Weinberg, an unashamed 'technological fixer'. Weinberg's view was coloured by his position on one side of the Cold War, looking across at what he regarded as failed and illiberal Soviet 'social engineering' (Weinberg 1966). Prefiguring a debate that would resurface with geoengineering, Weinberg suggested that technological fixes could be used to 'buy time' while society catches up with much-needed changes.

As we have grown more sceptical of grand technological schemes, the term has acquired pejorative connotations. But as Dan Sarewitz and Richard Nelson (2008) argue, we should find ways to separate the good from the bad.

There are technological fixes that appear to do an extraordinarily good job of using 'the power of technology in order to solve problems that are nontechnological in nature' (Volti 1995, p. 23), to adopt one definition. The example used by Sarewitz and Nelson is vaccines.

There are countless examples of bad technological fixes. People may of course disagree on such judgements, but we might point with relatively little controversy to schemes such as the 1980s' Strategic Defense Initiative, a proposed technological fix for the problem of nuclear war known popularly as 'Star Wars'. (This idea was advanced by arch-fixer Edward Teller, who would go on to develop a strong interest in geoengineering. Weinberg, writing in 1966, credited Teller with having already 'supplied the nearest thing to a Quick Technological Fix to the problem of war . . . The Hydrogen Bomb' [Weinberg 1966, p. 6]. A similar claim had been made by Orville Wright in 1917 about the aeroplane [Kelly 2010].) Sarewitz and Nelson choose the more mundane example of technologies for literacy, which have failed to improve upon conventional education despite decades of effort.

Separating good from bad fixes, according to Sarewitz and Nelson (2008), requires asking three questions: Does a proposed fix address the root cause of the problem? Can we tell when it's worked? Are the research and technology there already? Answering these questions may in some cases be straightforward. Most of the time, however, technological fixes will be ugly: they will sort out some problems while hiding, postponing or exacerbating others. The Green Revolution, for example, introduced new technologies for global agriculture after World War II. It dramatically increased crop yields for people in some parts of the world, avoiding acute hunger in India and elsewhere. But its successes did not extend to Sub-Saharan Africa, and it deepened a dependence on fertiliser and pesticide use that would prove unsustainable (Collingridge 1980; Royal Society 2009).

'Technological fix' is a misnomer. They are never fixes in the sense of repairing something that is broken. As Thomas Hughes explains, 'Technological fixes are partial, reductionist responses to complex problems. They are not solutions' (Hughes 2004, p. 241). Twenty-first century global problems are rarely acute. They are more often chronic, requiring management rather than a cure. Alvin Weinberg's faith in technological fixes now seems anachronistic. It is hard to imagine a contemporary government taking seriously schemes such as Star Wars. But, according to Evgeny Morozov, the enthusiasm for fixes is undimmed. He recasts it as 'solutionism', which 'presumes rather than investigates the problems that it is trying to solve' (Morozov 2013, p. 6). Morozov's focus is the new array of digital technologies for which claims are made that they will not just make our lives easier, but also solve big social problems. We see with other emerging technologies that similar claims are made about the future. Emerging technologies are often constructed as solutions to problems of their own imagination, as well as being imagined as problems for governance in themselves, too. Following Morozov, as well as the growing field of the sociology of expectations and an older set of ideas about the sociology of social problems (e.g. Schneider 1985), social scientists should therefore seek to track the claims that are made about the definition of such problems.

Responsibility, science and innovation

In response to the emergence of nanotechnology as an issue of public and scientific concern in the early 2000s, governments around the world rushed to consider whether new policy measures were warranted. The European Union (EU) decided that no new specific regulations were required but that the uncertainties of nanotechnology justified some form of early governance. Janez Potočnik, then commissioner for research, launched the EU Code of Conduct for Responsible Nanosciences and Nanotechnologies Research (NanoCode), which was drawn up in 2008 (EC 2008). It is a voluntary measure, aiming to give researchers and others a clear sense of what is expected of them and provide a basis for future policy discussions. At its heart is a set of seven principles: accessibility, sustainability, precaution, inclusiveness, excellence, innovation and accountability.

If it works, it could be a powerful way of helping members of the nano community to consider and organise their responsibilities (von Schomberg 2007). But nanotechnology is notoriously hard to demarcate. Nano research ranges across the sciences from materials science to pharmacology, and there is little agreement on definitions or implications. The reaction to the NanoCode among scientists, even those who would support its rationale, has been nervousness.

The nanoscientist Richard Jones, who has, in his promotion of public dialogue and scientific governance, acted as a conscience for the nascent nano community, highlights a concern held by many scientists. He is worried that responsibility will be understood in terms of retrospective accountability. He warns that scientists 'might be alarmed at the statement that "researchers and research organisations should remain accountable for the social, environmental and human health impacts that their N&N [nanoscience and nanotechnology] research may impose on present and future generations"' (Jones 2009, p. 336). Pointing to an assumed 'division of moral labour', he argues that

> scientists who make an original discovery may have little influence in the way it is commercialized. If there are adverse environmental or health impacts of some discovery in nanoscience, the primary responsibility must surely lie with those directly responsible for creating conditions in which people or ecosystems were exposed to the hazard, rather than the original discoverers. Perhaps it would be more helpful to think about the responsibilities of researchers in terms of a moral obligation to be reflective about possible consequences, to consider different viewpoints, and to warn about possible concerns.
>
> (Jones 2009, p. 336)

Even though 'accountability' is less strict than the legal idea of liability, the approach recommended by the EU NanoCode could, if interpreted strictly, lead to scientific paralysis. If the code is to be strengthened, as some have suggested, the slipperiness of the term will make it easy for some to reclassify their work as mundane chemistry rather than cutting-edge nanotechnology. (This was the outcome of the UK government's introduction of a voluntary reporting scheme for companies working on nano research; see Stilgoe 2007.)

Scientific research is an important part of the wider innovation system. While there is plenty of scientific activity that is not driven by the demand for innovation, for practising scientists, the imperative to innovate is proving harder to ignore. Gibbons *et al.* (1994) see this as a move from mode 1 science, which is defined by separation from its social context, to mode 2, in which the boundaries between science, society and industry blur. As science in many areas is increasingly connected with its 'context of application', questions of responsibility become harder to escape from. Science and innovation are, in this regard, different but crucially connected. We should therefore consider the responsibilities of science and scientists as part of a broader discussion of responsible innovation.

Understanding and governing the social and ethical dimensions of innovation are hugely complex. The unpredictability of technology's effects has in the past provided an excuse for inaction. Innovators have been expected to innovate, and society has been expected to live with the consequences. The externalities of innovation create a fundamentally unequal relationship between current and future generations. If the futures we enable through innovation end up better than the present, then that is a happy accident. But if they are worse, then future generations have no recompense for the burdens we have given them. If we recognise that emerging technologies are fateful, then innovation creates what philosopher Hans Jonas calls the 'imperative of responsibility' (Jonas 1985).

Innovation and responsibility are not easy bedfellows. Innovation is, according to Ulrich Beck (2000), a central feature of society's 'organised irresponsibility'.[7] There is a view held by some scientists that control of science is undesirable, if not impossible. Michael Polanyi argued that 'you can kill or mutilate the advance of science, you cannot shape it' (Polanyi 1962, p. 64). The web of responsibility that should ideally sit over the innovation system is often in reality cobweb-thin. It is easy for parts of the system to offload responsibility onto others.

The conventional story told by scientists throughout history is that they are responsible only to each other, as a community. Their job is the production of knowledge, and it is the responsibility of others to employ this knowledge. Science, according to this story, is free from values and ideals. This story no longer holds, if indeed it ever did. In most areas of research, and particularly those with innovation or policy implications, science is unavoidably entangled in applications, innovations and politics (Latour 2008). In reality, science and innovation are being shaped all the time by the people who do them and the network of organisations around them, and they can therefore be reshaped in more responsible ways (Guston 2008).

Scientists have throughout history actively sought responsibility in involving themselves, as experts and as citizens, in political and ethical issues. Especially since World War II, when science was implicated in what many considered undesirable applications, such as the atomic bomb, some scientists have come together in groups such as Pugwash, the Society for Social Responsibility in Science (and its British offshoot), the Union of Concerned Scientists, and Scientists for

Global Responsibility. These bodies provided forums for scientists to question the connections between science and the undesirable interests of companies and governments. At the same time, individual scientists have notably taken responsibility for reshaping the ways in which science is conducted and governed, as well as used, with prominent examples including Jonas Salk's and Tim Berners-Lee's rejection of patents for, respectively, the polio vaccine and the World Wide Web; and John Sulston's insistence that data from the Human Genome Project be quickly made public.

The growing power of humans to reshape their world has led to what Ulrich Beck (1992) has described as a 'risk society'. But while these new risks, whose impacts can be felt across continents and generations, seem to have prompted a new 'era of responsibility' (Strydom 1999; Pellizoni 2004), what this means in practice remains elusive.

Scientists' concern about the EU NanoCode can in part be explained with reference to different ideas of responsibility. For accountability to be accurately determined, it must be traced along lines of causality. With research and innovation, these lines are often impossible to draw, even in hindsight. Accountability assumes that the past and the present provide a reasonable guide to the future. Innovation acts to change the future and therefore redraws accountability (Adam and Groves 2011).

The time dimension of responsibility is crucial. Luigi Pellizoni offers a useful typology according to whether we are considering responsibility after the fact or before. Retrospective responsibility deals in concepts such as accountability, liability and blame. In matters of regulation and control, such concepts are vital, but their applicability to emerging science and innovation is unclear. We need a view of responsibility that points forward. Prospective responsibility, according to Pellizoni, is about care and responsiveness (Pellizoni 2004). Society expects a duty of *care* from parents towards their children, even though parents can neither know nor control the future lives of their children. We therefore expect that parents will be *responsive* to changing demands and changing information.

We might offer 'taking care of the future' as a slogan for responsible innovation, but the phrase 'taking care' is not self-evidently good for the governance of emerging technologies. It can imply humility or hubris, depending on its context. One need only consider the difference between a parent asking a nanny to 'take care' of someone and a mafia boss asking a trusted lieutenant to 'take care' of someone to see the difference. In the first case, the understanding of responsibility is rooted in a relationship, an attachment and a practice, an important point developed by discussions of care ethics. 'Caring for' something or someone is distinct from the more paternalistic idea of 'caring about' something or even 'taking care' of it, in the sense of sorting it out. For Annemarie Mol, writing in the context of healthcare, an emphasis on care is not about grand ethical principles, but rather about everyday practices: 'Care is a process: it does not have clear boundaries. It is open-ended . . . Care is not a product that changes hands, but a matter of various hands working together (over time) towards a result' (Mol 2008, p. 18).

In exploring questions of responsibility in and around science, we should not underestimate the cultural and structural pressures that make it hard for scientists to take responsibility, even if as individuals they are predisposed to care deeply about what they are doing. The institutions and cultures of science jealously guard the autonomy of scientists. The role responsibilities of scientists – to create robust knowledge – are seen as trumping the general responsibilities that scientists may have towards society (Douglas 2003). This creates an asymmetrical distribution of credit and accountability, or, as Jerry Ravetz puts it, 'Scientists take credit for penicillin, while Society takes the blame for the bomb' (Ravetz 1975, p. 46).

The demand for scientific autonomy has led scientists to argue for a 'right to research', particularly in controversial fields such as stem cell science in the USA (Benjamin 2013). However, as M. B. Brown and Guston (2009) explain, a 'right to research' need not be understood only in the negative sense of freedom *from* interference. There is also a republican view of rights that would see science as appropriately embedded within wider society rather than detached from it. The freedom *to* research would here be imagined in a more positive sense. The autonomy and purity of science are not a given; they are argued for and worked at by scientists and their institutions (Nordmann 2010).

Questioning the products, processes and purposes of innovation

Public, critical discussions about innovation typically raise a number of questions, which we can divide into *what* questions, *how* questions and *why* questions (Box 2.1). These questions have emerged as important from reviews of public dialogue exercises, and they have solid foundations in social science and philosophical analysis of emerging technologies as well (see, for example, Feenberg 1991; Macnaghten and Chilvers 2012). They are the questions that often come to define public controversies about new technologies.

Box 2.1 Questions of responsibility

***What* questions (about the products of science and innovation)**

- What are the likely benefits and to whom?
- What are the likely risks and to whom?
- What other impacts can we predict?
- Might these change in the future?
- What might the unintended consequences be?
- What don't we know about?
- What might we never know about?

(continued)

(continued)

How questions (about the processes of science and innovation)

- How should research and innovation take place?
- How should standards be drawn up and applied?
- How should risks and benefits be defined and measured?
- How might this technology be used in unintended ways?
- Who is in control?
- Who will decide?
- Who will take responsibility if things go wrong?
- What if you are wrong about impacts?

Why questions (about the purposes of science and innovation)

- Why should this research be undertaken?
- Why are you in particular doing this research?
- What future are you envisaging?
- What are the intended consequences?
- Who will benefit?
- What are the alternatives that we could be considering?

These three sets of questions – *what*, *how* and *why* – relate respectively to the products, processes and purposes of innovation. Responsible innovation means ensuring that these questions are put on the table. As scientific innovation has become more powerful, scientists and society have grown used to the idea that research should be governed to minimise risks. The assumption in the past has been that the legitimate public interest lies in the *products* of innovation – the positive and negative impacts of technologies and the changes they enact upon societies and economies. We also now appreciate that when we can't predict or account for these impacts in advance, it is also sometimes right for society to take an interest in the *processes* of science. We expect oversight of ongoing research involving humans, animals and tissues, such as stem cells, with clear social and ethical dimensions. Areas of research such as geoengineering, where intent is precisely the matter of concern (Stilgoe 2011), suggest there is a legitimate interest in scientists' *purposes*. If scientific research is a journey into the unknown, then the direction of travel should be considered to be as important as the mode of transport.

Dimensions of responsible innovation

Governing the processes and purposes of science and innovation, in addition to their outcomes, means rethinking the frameworks and policies that currently shape research. In previous papers, colleagues and I have proposed a framework for responsible innovation (Stilgoe *et al.* 2013) that draws attention to four dimensions of responsibility: anticipation, inclusion, reflexivity and responsiveness.

Anticipation

David Guston (2012) describes how, in 1945, the chemist and philosopher of science Michael Polanyi joined Bertrand Russell, another philosophical luminary of the day, on the BBC radio show *The Brains Trust*. The panel were asked whether they could anticipate any possible applications of Einstein's theory of relativity. Reflecting later on this episode, Polanyi described how neither of them could think of any, 'But actually, the technical application of relativity . . . was to be revealed within a few months by the explosion of the first atomic bomb' (Polanyi 1962, quoted in Guston 2012). Polanyi expressed some embarrassment at his lack of foresight, but his ignorance was at least partly intentional. Polanyi's philosophy of science fiercely resisted any attempts to control what he called 'the republic of science'. His refusal to predict allowed him to argue that 'Einstein could not possibly take these future consequences into account when he started on the problem which led to the discovery of relativity' (Polanyi 1962, quoted in Guston 2012).

But while two of the cleverest men of the day did not anticipate the atomic bomb, even while it was under construction, others could. Frederick Soddy, another leading UK chemist, had begun discussing the perilous potential of atomic energy before World War I, lending his expertise to H. G. Wells, among others. Soddy had a clear sense of a responsibility to consider 'possible technical uses' of his and his colleagues' work, even though these may have been highly uncertain and decades away from realisation. Guston's (2012) conclusion is that anticipatory governance is both possible and necessary (see also Guston 2008) but that scientific culture is prone to certain sorts of anticipation and not others.

Organisations with a clear role in defining the future, and therefore a responsibility to consider their role, can find themselves being constantly reactive. The detrimental implications of new technologies are often unforeseen, and early warnings of effects are systematically ignored (Harremoës *et al.* 2001). As Bruno Latour (2010, p. 486) puts it, innovators are habitually 'fleeing ahead looking backwards'. The lessons of past mistakes have led to legislative moves such as the precautionary principle, but precautionary practice has tended to exacerbate the focus on risks rather than benefits.

As well as mapping what is known and what is likely, anticipation forces thinking about unknowns, with the aim of increasing resilience in the face of inevitable surprises. Anticipation prompts researchers and organisations to ask 'what if . . . ?' It is necessary in order to reveal new opportunities for innovation as well as to construct agendas for socially robust risk research and risk management.

Contemporary emerging technologies are accompanied by a glut of expectations, but they are typically narrow. Research in the sociology of expectations (see Borup *et al.* 2006) suggests that science is increasingly forward-looking and that expectations themselves contribute to shaping technologies (see also van Lente and Rip 1998). Scientific researchers are often caught in an 'economy of promises' (Jones 2008) to which they feel compelled to contribute. The futures that they and those around them describe are not predictions, although they may be presented as such. They are instead promises of what can happen if science is given particular resources and freedoms (Geels and Smit 2000). These 'sociotechnical imaginaries' (Jasanoff and Kim 2009) are never purely technical. They are also imagined social worlds.

The futures of science and innovation are typically imagined as both inevitable and close. It is intriguing to witness what some have called the 'naturalisation of technological advance' (Felt and Wynne 2007) in phrases such as 'Moore's Law', which describes the exponential trend in computing power with the presumption that it will indefinitely and inescapably continue. Even futures that are highly speculative are imagined to be 'on the horizon' (Michael 2000) or 'just around the corner' (Evans *et al.* 2009). To justify research, investment and enthusiasm, innovations are presented as within easy reach. Discussions take place in what Bell and Dourish (2007, p. 134) call 'the proximate future'. It is relatively easy to see how such dynamics emerge. The development of emerging technologies is expensive and therefore necessarily 'shot through with strategic considerations' (van Lente and Rip 1998). The temptation for hype is therefore hard to resist, and the claims made for technology are typically most brazen at the earliest stages, when there is the least evidence to constrain their promises (N. Brown 2003). Critiquing or fleshing out such promises would mean engaging with the full complexities of sociotechnical systems, rather than offering a seductive technological shortcut.

We have seen from discussions around nuclear energy, the Human Genome Project and, more recently, nanotechnology that the discussions of visions and expectations that come to define motivation can become closed all too easily. Those domains have attracted an 'economy of hope' (Del Vecchio Good *et al.* 1990) that encourages unconfined promise: 'energy too cheap to meter', the 'book of life' (Nerlich *et al.* 2002) or 'complete control of the structure of matter'.[8] The sheer scale of the hype casts argument and scepticism as an unwarranted attack on ambition. So outlandish possibilities are allowed to evolve into probabilities, which are subsequently manifest in scientific and policy commitments with little questioning.

Scientists may be privately uncertain about the claims made on their behalf, but they are nevertheless implicated in the bolstering of the economy of promises. Scientists and the institutions of science may legitimately respond that hype is a necessary evil of contemporary science policy. Scientific megaprojects such as the Large Hadron Collider or the Human Genome Project would fail a policymaker's strict cost–benefit test, so a 'social bubble' (Gisler *et al.* 2011) of expectations can help to enrol policy and public audiences in a common project. Hype can, however, lure scientists and innovators into a 'Novelty trap' (Rayner 2004). The same novelties that are emphasised to attract attention and funding may reasonably arouse the interest of regulators. But the pattern has commonly been that when novel regulations are proposed, technologies are reclassified as mundane. The uniqueness and novelty of new technologies are more malleable than they appear.

At first sight, geoengineering's economy of promises would seem to be rather different from those of genetics or nuclear power, which have promised untold economic benefits from their innovations. Few people interested in geoengineering imagine that it will be an unalloyed good. Even enthusiasts for geoengineering admit that it would, as currently envisaged, create as well as solve problems. The dominant imaginary for geoengineering is one of potency rather than benefit. Geoengineering futures share a degree of millennial reflexivity with those of nanotechnology and synthetic biology, whose promises exist 'between extreme revolutionary potential on the one hand and despairing disappointment on the other' (N. Brown and Michael 2003 p. 4), which we could extend to include cataclysm. Synthetic biologist Drew Endy sees synthetic biology futures oscillating between utopia and dystopia in a 'half-pipe of doom' (Calvert *et al.* 2014).

As the Polanyi and Soddy story illustrates, anticipation does not depend on the strength of our crystal balls; it depends on whom we choose to talk and listen to. The aim is not to predict the future, but rather to think through various possibilities. For research funders, proximity to the research coalface brings an additional responsibility of helping other organisations anticipate the future. They are more likely to be involved in conversations of risks, benefits and opportunities that might only reach other decision-makers after it is too late. Making innovation more anticipatory therefore asks innovators to include new perspectives.

Inclusion

There is a growing recognition across a range of research areas that science is too important to be left to scientists alone. Responsible innovation should aim to include diverse perspectives and sources of expertise. In some areas, ideas of what counts as an innovator are broadening to include users, non-governmental organisations (NGOs) and other stakeholders. 'User-driven' and 'open' innovation is seen by writers such as Eric von Hippel (2005) and Henry Chesbrough (2003) as an important trend, although these authors are more interested in the speed and efficiency of innovation than its politics and direction.

Perhaps the clearest argument for greater inclusion comes from the wealth of experience in using science and technology for global development. Innovation is often sold on the promise that alongside its economic potential in the rich North, grand benefits will accrue to poor countries. In its assessment of the potential of nanotechnology, the Royal Society and Royal Academy of Engineering were scathing about such promises:

> Much of the 'visionary' literature . . . contains repeated claims about the major long-term impacts of nanotechnologies upon global society: for example, that it will provide cheap sustainable energy, environmental remediation, radical advances in medical diagnosis and treatment, more powerful

> IT capabilities, and improved consumer products . . . However, it is equally legitimate to ask who will benefit and, more crucially, who might lose out? . . . Concerns have been raised over the potential for nanotechnologies to intensify the gap between rich and poor countries because of their different capacities to develop and exploit nanotechnologies, leading to a so-called 'nano-divide'.
>
> (RS–RAEng 2004, p. 52)

The nano-divide originates from a set of unquestioned assumptions about the challenges faced by particular groups and the answers that technology might provide. It is an iteration of a familiar pattern. NGOs, researchers and funders have all tried to redress this imbalance, but much innovation still proceeds in ignorance of the problems that it claims to be addressing.

If a genuine geoengineering technology is to emerge at a planetary scale, it is likely to be ubiquitous and centrally controlled, even autocratic (Szerszynski *et al.* 2013). It is therefore unclear how anyone other than experts and governments might actively be involved in its development or use. The inclusion of other perspectives in geoengineering research and innovation will therefore necessarily entail other, more-proactive forms of upstream engagement. *The Economist* (2010) is right to conclude that 'producing plausible policies and ways for the public to have a say on them will be hard – harder, perhaps, than the practical problem of coming up with ways to suck up a bit of carbon or reduce incoming sunshine'.

The last decade has seen a proliferation of innovative new forms of engagement with people previously disconnected from science and technology. In environmental decision-making, this has in the past taken the form of stakeholder engagement on well-defined local issues. But this model has been broadened to encompass new forms of engagement, with members of the public. These small-group processes of public dialogue (or 'mini-publics' (Goodin and Dryzek 2006)) include consensus conferences, citizens' juries, deliberative mapping, deliberative polling and focus groups (see Renn *et al.* 1995; Chilvers 2010).

In some countries, such as the UK, Denmark and the Netherlands, there appears to be a degree of commitment to the idea of democratic discussion about science, but it has often taken place in a context of political confusion and crossed purposes (Macnaghten and Chilvers 2012). Critics have argued that there are substantial methodological limitations that only serve to highlight the high cost of staging public discussions (Horlick-Jones *et al.* 2007; Rothstein 2007; Marris and Rose 2010). The risk is that these 'technologies of elicitation' contribute to a new technocracy (Lezaun and Soneryd 2007; also Rose 1999). Reviewing these activities in the context of nanotechnology, the Royal Commission on Environmental Pollution argued that there is a need to move away from large, one-off public engagement exercises towards an ongoing system of gathering 'social intelligence' (Depledge *et al.* 2010).

Beginning with work on the Human Genome Project, emerging technologies have also seen new forms of engagement between the previously disparate disciplines of ethics, law, science and social science. Following criticisms that the Ethical, Legal and Social Implications (ELSI) Research Program of the Human Genome Project was too far removed from the core science, social scientists and scientists have more recently begun to collaborate more intimately. The Socio-Technical Integration Research (STIR) project in the USA has introduced social scientists into a number of nanoscience laboratories with the dual aim of studying science in action and trying to open up social and ethical reflection among scientists (Fisher 2007).

Reflexivity

Science should in principle have self-criticism at its core (Lynch 2000). It proceeds by questioning assumptions and is strengthened through scrutiny. But cultures and practices of science are less good at asking themselves difficult questions about their broader ramifications. The disorganised scepticism of public debate is often shut out in an effort to protect the 'organised scepticism' described and advocated by Robert Merton (1973). One feature of the new experiments in public dialogue on science and technology is that they have forced scientists, innovators and institutions towards greater reflexivity. Reflexivity involves holding a mirror up to one's own social, ethical and political assumptions and being mindful of commitments, aware of the limits of knowledge and conscious that a particular understanding of an issue may not be universal. Reflexivity becomes particularly important when scientific research engages with public and policy audiences.

In the context of science and innovation, where so much activity is driven from the bottom up by self-governing and self-motivated researchers, alternatives can be opened up by enhancing the 'reflections of natural scientists on the socio-ethical context of their work' (Schuurbiers 2011, p. 769). But this is not to say that reflexivity can be straightforwardly defined (Lynch 2000), nor is it the responsibility of individual scientists alone. There is a need for institutional (Wynne 1993), second-order reflexivity (Schuurbiers 2011), in which often-unquestioned values and assumptions that shape science, innovation and their governance are themselves scrutinised. Unlike the private, professional self-critique that scientists are used to, responsibility makes reflexivity a public matter (Wynne 2011a). Some areas of science policy, such as biobanking in the UK, are starting to see explicit recognition of the need for reflexive governance (Laurie 2011). It should be noted, however, that current incentive structures – research assessment, career progression, and peer review of grants and papers – press against public engagement and reflexivity (Royal Society 2006).

Responsiveness

There is a range of processes through which questions of responsible innovation can be posed. Responding to these questions has, however, been less straightforward. New ways of doing public engagement, technology assessment, scientific advice and foresight have led to more open, more engaged discussions of new technologies. Initiatives such as constructive technology assessment (Rip *et al.* 1995), real-time technology assessment (Guston and Sarewitz 2002) and upstream engagement (Wilsdon and Willis 2004) have prised open a space in which new discussions might take place, but their achievements have been partial and short-lived. Connections back to systems of innovation and policy-making themselves have been limited (Sciencewise 2010; Stilgoe *et al.* 2014).

Responsible innovation is necessarily responsive. Innovation can only claim to be responsible if it has the capacity to change shape or direction in response to public values. If not, then discussions are disingenuous. Responsiveness poses some big challenges for innovation, which is systemically stubborn. The history of technology tells us that innovation is 'path-dependent' and prone to 'lock-in', which can result in the triumph of bad technologies – the QWERTY keyboard being a classic example (David 1985). Changing direction is hard when corporate interests, political agendas, research capacities and technological standards serve to keep innovators on the same track. Scientists are likely to be personally and professionally invested in particular trajectories of research. Pellizoni (2004, p. 557) calls responsiveness 'an encompassing yet substantially neglected dimension of responsibility'. Its two aspects relate to the meanings of the word 'respond' – to answer and to react (Pellizoni 2004).

There is a tension in processes of science and innovation between tendencies to open up and close down options. Science habitually opens up new alternatives through questions, challenges and competition, but it can also close off options in the search for an ideal solution. A diversity of options, research portfolios and technologies can appear at first glance messy and inefficient. But, as in an ecosystem, diversity is an important feature of productive, resilient and adaptable innovation systems (Stirling 2007). Responsible innovation should not just welcome diversity; it should nurture it and scrutinise those patterns of governance, such as intellectual property regimes and technological standards that can act to close down innovation options. However, we should recognise that the 'de facto governance' (Kearnes and Rip 2009) of innovation is likely to follow 'a logic of unresponsiveness' (Pellizoni 2004, p. 558), meaning that responsibility is often dodged or postponed. Responsible innovation aims to make parts of this de facto governance more reflexive and more deliberate (following Fisher *et al.* 2006), such that opportunities for responsiveness can be made visible and acted upon.

Responsible innovation is a collective political responsibility (Grinbaum and Groves 2013) or co-responsibility (Mitcham 2003), shared by the various actors in the innovation system. Avoiding Beck's (2000) 'organised irresponsibility' means identifying 'second-order' responsibilities (Illies and Meijers 2009), 'meta-task' responsibilities (van den Hoven 1998) or 'meta-responsibilities' (Stahl 2013). These are the responsibilities to anticipate and gain knowledge of possible consequences and then to build capacity to respond to them to ensure that responsible choices can be made in the future. The first step towards responsible innovation is to make the system governable, described by some as responsibilisation (Shamir 2008; Dorbeck-Jung and Shelley-Egan 2013). In legal terms, we might differentiate between negligence, in which implications are unforeseen, and recklessness, in which implications are anticipated but disregarded (Douglas 2009). With innovation, we might ask for improved anticipation as part of innovators' due diligence. When it comes to emerging technologies such as geoengineering, we might therefore ask not just about the possible implications of technologies, but also, as ethicists have already begun to do (Gardiner 2010; Betz and Cacean 2012), about the responsibilities we have to future generations to enable them to make relatively open choices. However, in doing so, we should not presume certainty about ethical implications, as has been the case for much literature on emerging technologies that we might label 'speculative ethics'.

The trouble with speculative ethics

The front cover of the April 2013 issue of *National Geographic* magazine carried a picture of some animals that had been considered lost forever – a sabre-toothed tiger, a thylacine and a mammoth among others – emerging from a giant test tube (Zimmer 2013). The emerging technology under discussion was the use of reconstructed DNA for 'de-extinction'. The headline read 'Reviving extinct species: We can, but should we?' It is a headline that captures the spirit of, and the trouble with, what Alfred Nordmann has called 'speculative ethics' (Nordmann 2007). De-extinction is an idea extrapolated from laboratory techniques in cloning and synthetic biology research. Its popular appeal owes more to *Jurassic Park* than to any particular scientific breakthroughs. In 2009, researchers attempted to clone a Pyrenean ibex, a species that had become extinct just a few years earlier. An animal was born, via a goat that acted as its surrogate, but it died after just seven minutes because of organ failure, which is common in such clones. This has, so far, been the only animal brought back from extinction. We can conclude that we are some way away from the hyperbolic claim of the *National Geographic* that this is a technology 'we can' use.

Speculative ethics adopts without question particular imagined technological futures and extrapolates their ethical implications. Following the Human Genome Project's institutionalising of bioethics, we have seen each subsequent emerging technology acquire its own ethical analysis, from nanoethics (Allhoff 2007) to neuroethics (Levy 2007) and, most recently, 'big data ethics' (Richards and King 2013). The focus of Nordmann (2007) is on nanoethics, which adopts a typical posture of speculating on what might happen 'if' particular technologies are brought into being (also see Hedgecoe 2010). Social scientists, lawyers, philosophers and scientists have all joined the speculative project. In doing so, however, they reify the rhetoric that is their starting point. They contribute to a narrative of inevitability in which the technology and its ethics justify one another even while one is positioned as critique. What begins as a 'big if' quickly turns into a 'when'. In the meantime, a set of important short-term questions about how research should proceed is overlooked. The science itself is released from ethical scrutiny. One might legitimately argue that there is nothing intrinsic or unique to nanotechnology that means it deserves its own ethicists. Regardless, the growth of non-scientific analysis has been rapid. Looking at the patterns of referencing among social scientists around nano, Shapira *et al.* (2010) have shown that while they begin by citing the scientific research, they quickly become self-referential. We can already see a similar spiral of speculation around geoengineering. The scientists justify their research because others are taking it seriously as a set of technological or policy options, and the social scientists, lawyers and philosophers join the maelstrom.

The rapid growth of scientific and ethical interest in synthetic biology has attracted criticisms that here too a spiral of speculative ethics is underway. Social scientists have begun to analyse and critique the growth of speculative ethics around synthetic biology, which, in the guise of highlighting concerns, serves to reinforce the potency of that technology (Marris and Rose 2012; A. Balmer and Bulpin 2013). Speculative ethics also reinforces the division of moral labour (Jones 2009) that hampers responsible innovation. Scientists increasingly recognise that emerging technologies are deserving of non-scientific attention. But speculative ethics leaves the science conveniently untouched. A recent article by a software engineer is typical of this view:

> It's important that geeks and suits and wonks get together and talk about these things . . . because geeks like me can do stuff like this, we can make stuff work – it's not our job to figure out if it's right or not. We often don't know.
>
> (Jim Adler, quoted in Zumbach 2013)

Some in the geoengineering community have echoed this view. For example, carbon capture researcher Greg Rau told a geoengineering email forum that 'ethics, economics, and politics should enter the equation once research tells us if we actually have any technically and environmentally viable options'.[9]

For most other geoengineering researchers, this separation is not yet so starkly assumed, but there is no shortage of speculation. Ethics, whether it is deontological or consequentialist in flavour, tends to downplay the uncertainties of technological futures in its search for implications. As I discuss in Chapter 3, we are already seeing a mass of research on the consequences of geoengineering, inspired by ethics and economics. Even if we assume we can know these effects, their incommensurability undermines such calculations (Hamilton 2013, p. 160). David Keith's point that 'people aren't discussing apples and oranges, they are talking about apples and oranges and Porsches and whales and moons' (quoted in Tollefson 2010) applies not just to the diversity of geoengineering proposals, but also to the range of their possible ramifications.

Philosophers such as Michael Sandel take an alternative, communitarian approach to emerging technologies. In a prominent piece on human enhancement, Sandel argues that the 'fundamental question is not how to ensure equal access to enhancement but whether we should aspire to it in the first place' (Sandel 2004, p. 52). The trouble is that such arguments still rest on a shared understanding of what the 'it' of enhancement is. Given the uncertainties and ambiguities surrounding emerging technologies, ethicists can provide useful guidance, but they cannot predict the future shape of these technologies any more than scientists can. (A paper that verges on a caricature of speculative ethics offers a consideration of human enhancement for environmental ends [Liao *et al.* 2012]. Hamilton [2013] takes such thinking more seriously than I would, but I share his concern that it seems indicative of a detachment from sensible ethical questioning.)

As I argued earlier, there is a clear need for improved anticipation in science and innovation, but this should be done with the aim of prompting thinking about present responsibilities rather than deferred impacts (Guston 2014). The aim should be to prepare for the unexpected by scrutinising, rather than enabling, the possibility of 'taking hold of the future' (Nordmann 2010, p. 10). A more urgent task for ethics, in conversation with science, is to provide a reality check and new forms of accountability on promises for emerging technologies (Fortun 2005). Building an 'ethics of promising' (Fortun 2005) also means tackling the conditions that encourage such promises (Groves 2013; Simakova and Coenen 2013).

Governing collective experiments

Geoengineering, in common with other emerging technologies, is talked about as being 'risky' (Kintisch 2010). It is a commonplace to suggest that new technologies introduce new risks, but the idea of risk can never fully capture society's concerns with new technologies. Describing geoengineering as risky is like describing human cloning or nuclear war in the same terms, which hardly captures the social context of these technologies. In many cases, new risks are not comparable with the old ones. To give a topical example, the risks of a self-driving car are qualitatively different – even if Google calculates them as quantitatively smaller – from the clear, high, well-understood risks of a car driven by a flawed human being. Crucially, many of the risks of new technologies are by definition not well understood because of their novelty. We should therefore distinguish between risk and other types of uncertainty.

Alongside risk, we can place uncertainty, when hazards are known but probabilities are unknown; ambiguity, when there is disagreement on the nature of the hazards; and ignorance, when there is no knowledge on which to assess either hazards or probabilities (Stirling 2010, following Wynne 1992). The tendency is for these more profound areas of 'non-knowledge' to be overlooked as policymakers reduce governance questions to those of risk (Wynne 2005; Stirling 2010).

As with speculative ethics, conventional technology governance presupposes a strict division of labour between those who innovate and those who regulate. A constructive, more responsible form of innovation would bring these together in a form of 'collective experimentation', built on the recognition that innovators always in some respects govern themselves and that policymakers are also a source of innovation.

The word 'experiment', like the word 'innovation', has come to mean almost anything its users would like it to – a generic, progressive byword to sell a new activity, with the implied advantage that proponents need not be completely sure about the outcome. For our purposes, the meaning of the term can be tightened up a bit, but not so much that we restrict the flow of the argument later. An experiment, then, involves the deliberate use or observation of a system in which certain things are controlled in order to measure effects.

Historians of science have over the last few decades turned their attention back to experiments. Experimentation was central to the emergence of early modern science in the fifteenth and sixteenth centuries. Stories of Galileo and Newton are stories not just of great thinkers, but of experimenters engaging with the natural world in new ways, whether dropping things from towers or building powerful telescopes. Steven Shapin and Simon Schaffer (1985) have described the battle between Robert Boyle's experimentalism and Thomas Hobbes's natural philosophy in the seventeenth century as a disagreement over the credibility of different forms of knowledge. A focus on material experiments brings the story of science that is populated by weightless ideas, lone geniuses and their 'thought experiments' down to Earth. Experimentalists may strive for what Schaffer (1989) calls 'transparency', the idea that the experiment itself is seen as unimportant when compared to the reliability of knowledge it is seen to produce. It is for this reason that experiments are often airbrushed out of history, like the scaffolding removed at the end of a building project. A popular account of scientific discovery may spotlight a singular experiment or piece of apparatus, but the production of knowledge demands more extensive infrastructure. 'Experimental systems' are a way of reducing complexity and controlling uncertainty. They are manifestations of what is known but designed to generate new surprises (Rheinberger 1997). For Rheinberger, 'experimentation, as a machine for making the future, has to engender unexpected events' (pp. 32–33). Failure and error are accepted as part of the process, although an experienced experimenter will seek to control their bounds.

A scientific experiment should require some degree of stability, control and reproducibility. In practice, scientific experiments rarely live up to this ideal. Replications are never replicas (Collins 1992), and, particularly in medicine or ecology, where humans or ecosystems form part of the apparatus, scientists find experimental control hard to maintain (Radder 2009). The interplay between idea and experiment can take many forms. Some experiments are 'exploratory', detached from a clear hypothesis, with a greater possibility of surprise. Scientific experiments, according to Hacking (1983), do not have to be preceded by theories.

Experiments involve things – instruments, apparatus, reagents, laboratories – and demand a degree of craft. In this respect, they blur the line between science and technology. Historians have shown the ways in which scientific research is shaped by the experimental technologies that scientists have at hand (Gooding *et al.* 1989). Experimental technologies give substance to, and so sustain (or 'lock in'), particular ideas and forms of research.

As well as depending on technologies, experiments are technological in another important sense. They go beyond neutral observation towards manipulation, from 'representing' to 'intervening' in the material world (Hacking 1983). Passive observation and measurement can be contrasted with active experimentation using instruments whose purpose is to create new phenomena or reproduce natural phenomena such as electricity or a vacuum in the laboratory (Hackmann 1989). Habermas (quoted in Radder 2009) argues that experimentation turns science into 'anticipated technology'. Experiments involve the 'systematic production of novelty . . . making and displaying new worlds' (Pickstone 2000, pp. 13, 30).

Experiments, therefore, play a central role in technological innovation. Historical and philosophical studies of experiments suggest that their job is not just to generate knowledge, but also to cement the credibility of knowledge and communicate it beyond the immediate group of experimentalists. But the public role of experiments is problematic. They are typically private activities. Historians have to dig into laboratory notebooks and transcriptions to work out what really goes on inside experiments, while anthropologists spend time inside laboratories observing how the reality of experimental life differs from the tidied-up reports in scientific journals (e.g. Latour and Woolgar 1979). Shapin describes how, at the time of Boyle, Hooke and the other experimentalists who would go on to constitute the Royal Society, there was a distinction between 'trying' an experiment, 'showing' an experiment and 'discoursing' upon an experiment. The trying – what we might consider genuine experimentation – allowed for the possibility of surprise as part of an effort to get the experiment to 'work' in a repeatable, reliable way. Only once this was achieved could the experiment be shown and talked about in public (Shapin 1988a). Harry Collins (1988) gives a twentieth century case of a public experiment in which a train is crashed into a nuclear fuel flask to demonstrate the safety of the latter. The institutions involved were understandably unwilling to take risks with the outcome of such an experiment or with the clear public message it was designed to communicate. This 'experiment' was in reality a public display. The experimenters maintained a strict control over the uncertainties involved.

The second half of the twentieth century saw growing recognition that scientific experiments could not be considered purely private. Since World War II the governance of scientific research has come to reflect social and ethical concerns with experimentation involving humans and animals. Public outrage over Nazi medical research and the Tuskegee Syphilis Study led to codes of research ethics – the Nuremberg Code (Nuremberg Military Tribunals 1949) and the Belmont Report (NCPHSBBR 1979), respectively – to protect the subjects of scientific research. Research involving embryos, genes and stem cells has deepened the ethical debate. During the same period, there was growing recognition of the environmental and health hazards of industrialisation. Precautionary regulations (see Harremoës *et al.* 2001) have sought to prevent people from becoming unwitting experimental subjects for uncertain foods and drugs.

Technology as social experiment

Despite scientific attempts to maintain control of the idea and practice of experimentation, experiments and experimental talk have become a common part of debates about the democratisation of science and innovation. Interest groups, expressing a sense of dehumanising powerlessness in the face of contemporary innovation, may invoke the idea that the public have become unwitting 'guinea pigs' in a form of experiment. A recent French documentary has the title *Tous Cobayes? (Are We All Guinea Pigs?)* (Jaud 2012). Its poster depicts the two technologies that have become the focus of analysis and critique, particularly in Europe – nuclear power and genetically modified crops. In some cases, groups have sought to reclaim experimentation, even assuming the guinea pig label with pride (Weinstein 2001). But for most, the language of experiment is an expression of disenchantment with technology. John Gray (1999) sums it up like this: 'The world today is a vast unsupervised laboratory, in which a multitude of experiments are simultaneously under way. Many of these experiments are not recognised as such.'

Much of the scholarly exploration of technology as a 'social experiment' was developed with reference to nuclear power stations. In 1972, technological fixer and nuclear enthusiast Alvin Weinberg admitted the limits of risk assessments for nuclear power:

> There is no proof that every conceivable mode of failure has been identified. Because the probability [of a nuclear accident] is so small, there is no practical possibility of determining this failure rate directly – i.e., by building, let us say, 1,000 reactors, operating them for 10,000 years and tabulating their operating histories.
>
> (Weinberg 1972, p. 211)

A few years later, a major accident at the Three Mile Island nuclear power station in Pennsylvania demonstrated the incompleteness of prediction, containment and control. Technology, in some cases, is 'unruly' (Wynne 1988); it simply does not do what it is told. In the decades since, there have been major accidents at Chernobyl and, most recently, Fukushima. These three 'data points' (to which we might add the 1957 Windscale fire; Wynne 2011b) would seem to constitute a rather small dataset, and each tells us different things about the problems of nuclear risk assessment (Pfotenhauer *et al.* 2012). One could add that the human cost of these accidents has been dwarfed by other technological disasters, such as the bursting of the Banqiao Dam in Henan, China, in 1975, which killed tens of thousands of people.[10] But as Krohn and Weingart (1987, p. 52) describe, 'Each nuclear power plant is its own test case. It may be categorized as an implicit experiment, and its most revealing case is the accident.' Nuclear power plants are sold as safe, but the complexity of the full sociotechnical system makes complete safety impossible. It is in this sense that Charles Perrow (1984) describes such accidents as 'normal'. Although each accident may have unique details, we can point to shared systemic failings that characterise such incidents (see Perrow 1984).

Sheila Jasanoff looks back on Perrow's (1984) book and reveals its astonishing prescience. In the three years following its publication, we saw disasters involving a chemical plant in Bhopal, a nuclear power station in Chernobyl, the *Challenger* space shuttle and the discovery of a new cattle disease, bovine spongiform encephalopathy, that would force a rethink of British science policy. According to Jasanoff (2003, p. 223), these crises 'have served collective notice that human pretensions of control over technological systems need serious re-examination'. She explains that

> risk . . . is not a matter of simple probabilities, to be rationally calculated by experts and avoided in accordance with the cold arithmetic of cost–benefit analysis. Rather, it is part of the modern human condition, woven into the very fabric of progress. The problem we urgently face is how to live democratically and at peace with the knowledge that our societies are inevitably 'at risk'.
>
> (Jasanoff 2003, p. 224)

Ulrich Beck puts it like this: 'given the indeterminateness of risk, existential experimentalism is unavoidable' (Beck 2009, p. 5).

The financial crisis of 2008, about which Perrow (2010) also wrote, reminds us that not only have our sociotechnical systems continued to increase in complexity and opacity but our institutions have got no better at comprehending the uncertainties that dramatically reveal themselves. The nuclear accident at Fukushima led Perrow (2011, p. 52) to conclude that 'some complex systems with catastrophic potential are just too dangerous to exist, not because we do not want to make them safe, but because, as so much experience has shown, we simply cannot'.

Clearly, not everyone would agree with Perrow's (2011) recommendation. Many would argue that the risks of nuclear power are socially acceptable, even if their calculation is problematic. But his conclusion that safety is illusory remains solid. Risk should never be the only criterion for assessing technologies. As Rayner and Cantor (1987, p. 3) put it, 'While assessments of probabilities and magnitudes of undesired outcomes are essential to making engineering decisions about competing designs or alternative materials, they are largely irrelevant to societal technology choices.'

The experimentality of complex technological systems and their regulatory partners suggests a new way of thinking about governance, focussing not on risk but on the conditions for legitimate experimentation (van de Poel 2011). We might ask some of the following questions about technology-as-experiment: Who are the experimenters? Who are the subjects? Where does the experiment begin and end? What are the certainties? What are the uncertainties? What surprises is the experiment designed to generate? What about other surprises? What's under control (in both the scientific and political senses)? How can it be stopped or reversed?

Recognising technologies as experimental should also demand ongoing monitoring and data collection, as well as capacity building to learn from and respond to new findings. Such 'technologies of humility' (Jasanoff 2003) should demystify technology, forcing innovators to confront 'the questions we should ask of almost every human enterprise that intends to alter society: what is the purpose; who will be hurt; who benefits; and how can we know?' (Jasanoff 2003, p. 240). In some cases, where the stakes are known to be high and experience contains some hard lessons, some aspects of social experimentality are formally recognised in governance. With pharmaceuticals, for example, processes for credible experimentation through phases of animal and clinical trials are well established. Following controversies over thalidomide and other drugs with side effects in particular subgroups, monitoring of medicines after they have been prescribed was stepped up. In the UK, the Yellow Card scheme has been a 'technology of humility' for the reporting of drug dangers.[11] This was at first the privilege only of medical professionals but has now been extended to patients themselves (Stilgoe *et al.* 2006).

With genetically modified crops, the conditions for experimentation have been bitterly contested. The development and testing of crops in conditions that many scientists regarded as controlled was challenged by NGOs on the basis that such experiments could never be hermetically sealed from either the natural or political environments in which they were conducted. In a form of extended 'experimenter's regress' (Collins 1992) field- and farm-scale trials that were expected to settle the debate only contributed to the controversy. Trials of genetically modified crops 'entrenched' a particular model of experimentation, only for public controversy to radically reframe the nature and purpose of the experiments, forcing additional responsibilities onto the experimenters (Levidow and Carr 2007; Bonneuil *et al.* 2008).

Brian Balmer (2004) narrates how, in a culture of Cold War secrecy, human bystanders to a biological weapons experiment on animals became unwitting participants in it, redrawing the experimental system. Scientists' loss of experimental control can be positive. Isabelle Stengers (1997) describes how participants in experimental psychology experiments may not be as pure and passive as expected. For Stengers, a 'good' experiment, in order to learn new things, should allow its subjects some room to misbehave.

Experimental government

The experiments that characterise corporate research and development (Thomke *et al.* 1998) have in some areas become more visible and more explicit as they leak out of the laboratory. Online services are subject to continual innovation. They remain 'in development' even while being in use. Google routinely runs experiments with its services, providing tweaked versions of websites to samples of users to gauge responses. Google also offers this 'A/B testing' as a service to other websites. In June 2014, Facebook ran such an experiment in which its users unknowingly had their news feeds tuned to reflect more or less positive moods to gauge what the researchers called 'emotional contagion'. This experiment, unlike most corporate research and development, was reported in the open scientific literature, allowing wider scrutiny of its ethical basis (Kramer *et al.* 2014).

There is little sign that such controversies will impede the growth of corporate experimentation.

In the public sector, meanwhile, experimentation is starting to be embraced by technocrats who have argued for 'evidence-based policy' but seen a proliferation of value-laden 'policy-based evidence' (Stilgoe *et al.* 2006). The idea of extending scientific experiments into policy has attracted attention in the UK in recent years (Haynes *et al.* 2012). Some see this as a way to expand or purify the evidence base for policy (John 2013); others see the inherent unpredictability of a 'predict and provide' model (Owens 1995). This argument has a rich tradition. Some political scientists have called for trial-and-error policymaking (e.g. Wildavsky 1988) rather than wait-and-see precaution. This argument follows the line of social scientists who claim that because society and policy are themselves an experiment, social research and policy should themselves be deliberately experimental.

The Chicago School sociologist Robert Park (1929) regarded society, and Chicago in particular, as a laboratory. Sociologists such as Park recognised that pure experimental control was normally impossible and may, in cases where beneficial treatment or services are withheld, be unethical. Park's predecessor in the Chicago School, Albion Small, wrote that 'the radical difference is that the laboratory scientists can arrange their own experiments while we social scientists for the most part have our experiments arranged for us' (quoted in Gross and Krohn 1995). Donald Campbell, in making the case for an 'experimenting society' (Campbell 1969), further clarified the methodological implications of 'quasi-experimental' research (Campbell and Stanley 1966). As Noortje Marres (2012) describes, advocacy for experimental (as a byword for 'progressive') forms of society has a strong tradition in political thought dating back at least to John Stuart Mill.

One evangelist for both private and public sector experimentation has argued that just as medical science proceeds with incomplete understanding and incomplete predictive power through randomised control trials, so social science should go the same way through 'randomised field trials' (Manzi 2012). Melinda Cooper makes the point that *in vitro* biological experiments will always be incomplete. A new pharmacological compound 'needs to be ingested, metabolized and lived-with, in order to prove its eventfulness' (Cooper 2007, p. 2). An *in vivo* experiment will reveal things and convince people that a test tube cannot. Manzi adds that the messy apparatus of social life multiplies the problems of prediction, so we should engage with uncertainty experimentally. We need not fully understand the cause if we know the outcome we are looking for. Some areas of policy, such as international development, are already executing an experimental turn (see, for one prominent example, Duflo *et al.* 2013). Perhaps predictably, there have been suggestions that innovation policy should itself become more scientific through improved policy experiments (Bakhshi *et al.* 2011).

This trend could be taken to reflect humility in the face of social and political uncertainty. But it could equally be a new technocracy, the next step in the search for optimal, evidence-based policy. The enthusiasm for experimental government hides a spectrum of opinion and a mass of tensions. There are those who argue for randomised control trials as the ultimate indicator of 'what works'. Their aim is to narrow the control experiments in a search for the determinants of improved policy outcomes. Then there are those who argue, much as Wildavsky (1988) does, for truly adaptive governance as a process of ongoing learning. Recent statements from Silicon Valley CEOs Peter Thiel and Larry Page reveal a desire for an extreme libertarian version of experimental government inside unregulated communities (based at sea in Thiel's case).[12]

Experiments operate through the control of certain variables. As we will see later in this book, these variables become complex as humans and natural environments are brought into the system. But there is another sense of 'control' exerted by experimentation that is no less important. The important question is who has the power to frame, design and learn from the experiment.

Democratising experiments

Experiments can be used to close down debates, within and around science, as well as open them up. We have seen with recent moves to broaden public dialogue on issues involving science that even these 'technologies of democracy' (Laurent 2011) can be more or less technocratic, depending on the bounds set for the experimental system (Stilgoe *et al.* 2014). Alexander Bogner (2012) points to the paradox that while the social experiment of innovation becomes more visible, the practice of public participation often seems to take place under strict laboratory conditions.

Nevertheless, a number of scholars argue that we need to turn the language of experimentation back on science itself. A group of prominent STS researchers followed Bruno Latour (1998) in calling for 'collective experimentation' (Felt and Wynne 2007) as an alternative to the 'regime of technoscientific promises' that currently characterises the governance of science and innovation. If 'an experiment is a question' (Rosenblueth and Weiner 1945, p. 316), then collective experimentation is about democratising the asking and answering of the question. According to Helga Nowotny (2005), non-scientists have conventionally been presented with the results of experiments as though they were customers in a restaurant. Now, she concludes, 'we have to let the public also into the kitchen' (Nowotny 2005, p. 20).

Collective experimentation means rethinking the scope of the experimental system. If we recognise that people beyond science are involved in and affected by both scientific experiments and the broader experiments of technological innovation, we might argue for the inclusion of their perspectives in experimental design. If we are not quite at a situation where the whole world has become a laboratory (Latour 1999), we should at least question the temporal and spatial boundaries of experimentation (Davies 2010). Collective experimentation would value diversity and criticism as a source of resilience and variety, rather than dismissing dissent as 'anti-science', as is often the case in technological controversies (Stirling 2010). Transposed onto university research, collective experimentation also implies the mixing of disciplines previously considered separate, if not in competition. This book is the result of one process of mixing in which I have been involved, which has challenged some of the methods and conventions of social science.

Experimental methods

The conclusions I draw in this book are not based on neutral, detached observation. Over the four or so years that I have been involved with the geoengineering research community, I have participated in countless collective discussions, and my aim has been to shape and inform these discussions, as well as observe them. My research has therefore been an open-ended experiment in itself.

My involvement in geoengineering began as a staff member of the Royal Society's Science Policy Centre. Though my major responsibilities lay in other projects, I sensed and began to share some of the growing fascination and ambivalence of Royal Society fellows and staff as the institution conducted its study and prepared its report. My job was to be a policy adviser, not an ethnographer, so I have sought to reconstruct this story using interviews and public documents, which I report on in Chapter 4. After three years at the Royal Society, I became a researcher on a joint Economic and Social Research Council (ESRC)/ Engineering and Physical Sciences Research Council (EPSRC) project to develop a 'framework for responsible innovation' that could be used by EPSRC to better understand and govern emerging technologies. This project, created by Richard Owen from Exeter University and Phil Macnaghten from Durham, was prompted by concerns about synthetic biology, which were at the forefront of research policymakers' minds in 2011. While I was working with EPSRC, the Stratospheric Particle Injection for Climate Engineering (SPICE) project, which I had been dimly aware of while at the Royal Society, began to cause some concern to its funders. SPICE became a test case for the responsible innovation framework that we were developing. EPSRC agreed that as part of their oversight of the SPICE project I should extend my remit to explore stakeholder opinion about its open-air experiment. A further grant from the ESRC allowed me to deepen my collaboration with SPICE and to focus on the geoengineering researchers' own concerns, with the explicit aim of learning lessons from SPICE.

This accidental and gradual attachment to the world of geoengineering research raises methodological dilemmas that are becoming more familiar to STS researchers working in areas of emerging science and technology, despite being an older issue in sociology (Elias 1956). The STS discipline studies experiments, but it is also itself experimental. Enlightened scientists and policymakers have recognised that STS could be part of a renewal of science's relationship with its publics. In the twenty-first century, people trained in STS often find themselves 'embedded', alongside ethicists, industrialists and others, in various scientific efforts: research projects, advisory committees or teaching in university science departments. STS people have been granted unprecedented access, but as with embedded war reporters, this entails compromises (and prompts a suspicion that we are being kept away from the real action).

Some STS researchers see their role as studied detachment from their object of study (scientists), while others examine, and occasionally pull, the strings connecting science to society, policy and politics. Using Steve Fuller's (1993) distinction, subsequently developed by Sismondo (2008), we can say that the 'high church' of STS, tending to emphasise epistemology and focusing on esoteric sciences, has ceded ground to the 'low church', a group that is less ashamed of engaging in political debates. The UK, once the Vatican of high-church STS with its Strong Programme and Empirical Programme of Relativism, now has an STS congregation that is enthusiastically (if not evangelically) engaged in public and policy discussions about science, and the life sciences in particular. As Sismondo argues, these theological dividing lines have been overdrawn. Most STS people, despite their critical instincts, seem now to be relatively comfortable practising what Arie Rip (1994) calls 'constructive constructivism' (see also Newby 1992).

This is not to say that the relationship is easy. According to Brian Wynne (2006), the roles social scientists have been asked to play typically fall into two categories. Either we are asked to help mediate the relationship between scientists and non-scientists, or we are asked to predict and explain the social implications of new technologies – either facilitators or soothsayers. In either case, to return to the experimental metaphor, social scientists are usually cast as lab technicians rather than as paid-up members of the research team. The growing desire among scientists to explore the ELSI aspects of their work has prompted a critique of the 'ELSIfication' of social science (Guston 2004).

One prominent martyr to ELSIfication has been the anthropologist Paul Rabinow, who joined a major multidisciplinary synthetic biology centre to lead its human practices strand. During his short tenure within this new set-up, Rabinow described his intentions and experiences in some depth. He saw his job as working with the scientists to 'pose and repose the question of the ways in which synthetic biology is contributing or failing to contribute to the promised near future' (Rabinow 2009, p. 304). His aim was to create new possibilities, to be a part of the innovation project itself, but others in the centre saw his role as predicting and mitigating the risks of synthetic biology. The disconnect here between the expectations of the social scientist and those of the scientists is stark.

Rabinow and his colleague Gaymon Bennett recount their personal and epistemological experiences with the language of experiment. They talk about experimental 'design' and 'equipment', concluding that they 'have been practising science in the broad sense' and their experiment 'must be considered a success' (Rabinow and Bennett 2012, p. 1), even though their experiment found that true collaboration was all but impossible given the cultures of science they were working with. (One might question wonder whether they, as part of the apparatus, are able to adequately assess their own experiment, but this is a version of a common critique of ethnography.)

Nevertheless, there is much to learn from their experiment. Rabinow and Bennett draw attention to the distinction between cooperation, which employs a conventional division of labour, and collaboration, which 'anticipates the likely reworking of existing modes of reasoning and intervention' (Rabinow and

Bennett 2012, p. 6). Methodologically, this collaborative mode introduces myriad complications, not least the morphing of social scientists from merely observers or 'participant observers' to 'observing participants'. This 'anthropology of the contemporary' is unavoidably experimental (Rabinow and Bennett 2012, p. 46).

Rabinow and Bennett are critical of the conventional casting of ethicists in relation to science. They challenge the ways in which ethicists are expected to establish 'moral "bright lines"' (Rabinow and Bennett 2012, p. 4). As I discuss in this book's conclusion, the need to move beyond this view of ethics is a central part of the governance of emerging science and technology. My research therefore joins the efforts of others, many of whom are working in and around synthetic biology, to develop a post-ELSI social science (see also A. Balmer *et al.* 2012) in which social scientists are part of the project of knowledge production rather than setting the rules for scientists to follow.

STS has perhaps taken its maxim of 'following scientists around' (Shapin 1988b) too far when it comes to issues of public concern. With nanotechnology, it fell to an STS professor to tell the US Congress,

> I would not advise you to pass a Nanoethicist Full Employment Act, sponsoring the creation of a new profession. Although the new academic research in this area would be of some value, there is also a tendency for those who conduct research about the ethical dimensions of emerging technology to gravitate toward the more comfortable, even trivial questions involved, avoiding issues that might become a focus of conflict.
>
> (Langdon Winner, quoted in Fortun 2005)

By asking Congress to think differently about the role of social science in large scientific projects, Langdon Winner is also asking his fellow social scientists to think big, to not get stuck in scientific tramlines. But Winner's argument is for studied distance, whereas I would agree with Mike Fortun (2005) that 'cosiness' in relationships between social scientists and natural scientists can be both valuable and productive, particularly in 'promissory sciences like genomics, where the future is volatile and emergent'. Fortun (2005, p. 160) argues that STS should release its instincts of 'suspicion, antagonism, opposition, conflict [and] distrust' (see also Guston 2014). I see STS researchers at their most useful when they are focussing not on science as knowledge, but as experiment, with the experiment in question being as much social as technical. In high-profile new areas of science, where the ambitions are enormous and claims are hyperbolic, the experiment also encompasses the reorganisation of science. The machinery of governance, funding and regulation becomes part of the apparatus to be studied.

If geoengineering research is to proceed in a way that might be regarded as socially robust, we need to think very differently about it. As I describe in Chapter 5, over the last couple of years, geoengineering experimenters have understandably sought to define a space in which they might operate with relative autonomy. They are interested in collaboration with social scientists and ethicists inasmuch as they are keen to know where the bright lines may be drawn. But the risk is that, as we have seen with other emerging technologies, discussions that are initially open can close rapidly as money and researchers pour in. With geoengineering, STS has helped prise open an opportunity for the reimagining of science policy. An important experimental objective is to keep this space open.

My relationship with SPICE intensified over the life of the project. I began by collaborating with social scientists whose responsibility it was, on behalf of the project's funders, to construct a new governance mechanism (the 'stage-gate' described in Chapter 5), which placed a substantial additional burden on the project team. The SPICE principal investigator, Matt Watson, had every right to be frustrated and defensive, but he chose instead to regard it as an opportunity for a new form of experimentation. As I began to interview stakeholders, the SPICE scientists and other geoengineering researchers, my presence at times met with confusion, but I felt as though I was moving from being an observer of SPICE to an appendage of SPICE and then to a member of SPICE. Geoengineering researchers are used to having social scientists hanging around. But some of them were bemused that I chose to stay for scientific discussions rather than just discussions of 'geoengineering governance'. My response was that I was interested in the scientific bits of geoengineering rather than just the social. To their credit, most of the scientists took little issue once they had realised they had become the objects of my study.

It has helped that SPICE itself was a mixed bag of scientists, with almost a full range of views on the merits of geoengineering. I became SPICE's critical friend, or friendly critic, to better configure my approach and my job, legitimating discussions that were already happening within the team as much as introducing new ones. I began, as social scientists often do in the company of scientists, with an assumption that I was relatively powerless. However, as SPICE's technical experiment was revealed to be unavoidably social, my interviews and conversations became an exchange of views rather than a conventional social scientific extraction of views. When I began to write and speak about SPICE in public, the SPICE team rightly held me to account for my errors and misunderstandings. They asked whether my social science was playing by the rules I was proposing for responsible science and innovation. It became clear that I hadn't clarified my own aims or expectations to the scientists as our relationship had developed. I would agree with one SPICE scientist who described how this messy situation had produced an interesting outcome:

> *'The classic complaint I hear from social scientists is that scientists think social scientists are just there to communicate the research and that's certainly not been our relationship with social scientists . . . our relationship with the social scientists involved in the project will change over time . . . I think if you'd started off with a classical relationship between social scientists and scientists, I think you might have ended up in a different place, whereas I think this way round actually you will probably eventually end up in the perfect place, which is people truly thinking in an interdisciplinary fashion.'*

Although it starts with SPICE, my research speaks beyond the particular project to the small but growing world of geoengineering research, and beyond that to offer insights into politicised science and emerging technologies in general. But my assumption throughout is that it is necessary to engage with the practice of scientific research rather than just abstract geoengineering hopes and fears. Geoengineering is an area where social scientists have no monopoly on critical reflection, just as science has no monopoly on enthusiasm. This is not an area in which science can be juxtaposed against society in any straightforward way. There is more disagreement among scientists than between scientists and the outside world, but the quiet ambivalence of some researchers is often drowned out by the louder certainties of their more prominent colleagues. As my research has progressed, my main objective has moved from one of bringing critique to geoengineering research towards one of drawing out the critiques within geoengineering research.

In the chapters that follow, I describe some of my observations – I hesitate to call them experimental results – with a focus on the experimentality of geoengineering. My aim is to keep open, and in some cases to prise open, understandings of what geoengineering is, what constitutes an acceptable geoengineering experiment, and who should and shouldn't be involved.

Notes

1 The details of this story come from Jerry Garrett (2008).
2 Data from the US Department of Energy, Vehicle Technologies Office. Available online at http://www1.eere.energy.gov/vehiclesandfuels/facts/2009_fotw577.html (accessed 18 July 2013).
3 From the introduction to the novel *Crash* (Ballard 1995).
4 Figure taken from the World Health Organization. Available online at http://www.who.int/gho/road_safety/mortality/traffic_deaths_number/en/ (accessed 8 July 2014).
5 Latour and Venn (2002) give the example of a desk whose drawers can't be opened simultaneously. Anyone who has had a full filing cabinet tip over might understand why. But in the context of Latour and Venn's particular usage it remains a mystery.
6 See, for example, the Lund Declaration (Svedin 2009).
7 Barbara Adam and Chris Groves (2011) use the phrase 'structural irresponsibility'.
8 Eric Drexler, Foresight Institute briefing, 1988.
9 Greg Rau, email posted to the Geoengineering Google Group, 14 May 2013.
10 This case is mentioned in 'Geek Power?', a blog post by Richard Jones (2012).
11 http://yellowcard.mhra.gov.uk (accessed 10 July 2014).
12 Larry Page told a Google conference in 2013 that 'there's many, many exciting and important things you could do that you just can't do because they're illegal, or they're not allowed by regulation, and that makes sense, we don't want the world to change too fast. Maybe we should set aside a small part of the world . . . ' (Yarow 2013). Meanwhile, the director of Thiel's Seasteading Institute says their aim 'is to open a frontier for experimenting with new ideas for government' (Miles 2011, p. 2).

References

Adam, B., and Groves, C. (2011). Futures tended: care and future-oriented responsibility. *Bulletin of Science, Technology & Society*, 31(1), 17–27.

Allhoff, F. (2007). On the autonomy and justification of nanoethics. *NanoEthics*, 1(3), 185–210.

Bakhshi, H., Freeman, A., and Potts, J. (2011). *State of uncertainty: innovation policy through experimentation*. Provocation 14. London: NESTA.

Ballard, J. G. (1995). Introduction to *Crash*. London: Vintage.

Balmer, A., and Bulpin, K. (2013). Left to their own devices: post-ELSI, ethical equipment and the International Genetically Engineered Machine (iGEM) Competition. *BioSocieties*, 8(3), 311–335.

Balmer, A., Bulpin, K., Calvert, J., Kearnes, M., Mackenzie, A., *et al.* (2012). Towards a manifesto for experimental collaborations between social and natural scientists. *Experimental Collaborations* blog. Available online at http://experimentalcollaborations.wordpress.com (accessed 8 July 2014).

Balmer, B. (2004). How does an accident become an experiment? Secret science and the exposure of the public to biological warfare agents. *Science as Culture*, 13(2), 197–228.

Beck, U. (1992). *Risk society – towards a new modernity*. London: Sage.

——— (2000). Risk society revisited: theory, politics and research programmes. In B. Adam, U. Beck and J. van Loon (eds), *The risk society and beyond: critical issues for social theory* (pp. 211–229). London: Sage.

——— (2009). *World at risk*. Cambridge: Polity Press.

Bell, G., and Dourish, P. (2007). Yesterday's tomorrows: notes on ubiquitous computing's dominant vision. *Personal and Ubiquitous Computing*, 11(2), 133–143.

Benjamin, R. (2013). *People's science: bodies and rights on the stem cell frontier*. Stanford, CA: Stanford University Press.

Betz, G., and Cacean, S. (2012). *Ethical aspects of climate engineering*. Karlsruhe, Germany: KIT Scientific Publishing.

Bijker, W. E., and Law, J. (1992). *Shaping technology/building society: studies in sociotechnical change*. Cambridge, MA, and London: MIT Press.

Bogner, A. (2012). The paradox of participation experiments. *Science, Technology, & Human Values*, 37(5), 506–527.

Bonneuil, C., Joly, P-B., and Marris, C. (2008). Disentrenching experiment: the construction of GM-crop field trials as a social problem. *Science, Technology, & Human Values*, 33(2), 201–229.

Borup, M., Brown, N., Konrad, K., and van Lente, H. (2006). The sociology of expectations in science and technology. *Technology Analysis & Strategic Management*, 18(3/4), 285–298.

Brown, M. B. (2007). Can technologies represent their publics? *Technology in Society*, 29(3), 327–338.

Brown, M. B., and Guston, D. H. (2009). Science, democracy, and the right to research. *Science and Engineering Ethics*, 15(3), 351–366.

Brown, N. (2003). Hope against hype – accountability in biopasts, presents and futures. *Science Studies*, 16, 3–21.

Brown, N., and Michael, M. (2003). A sociology of expectations: retrospecting prospects and prospecting retrospects. *Technology Analysis & Strategic Management*, 15(1), 3–18.

Buchanan, C. (1963). *Traffic in towns: the Buchanan report*. London: HMSO.

Calvert, J., Schyfter, P., Elfick, A., and Endy, D. (2014). *Synthetic aesthetics: investigating synthetic biology's designs on Nature*. Cambridge, MA: MIT Press.

Campbell, D. T. (1969). Reforms as experiments. *American Psychologist*, 24(4), 409–429.
Campbell, D. T., and Stanley, J. (1966). *Experimental and quasi-experimental designs for research*. Boston: Houghton Mifflin.
Chesbrough, H. (2003). *Open innovation: the new imperative for creating and profiting from technology*. Boston: Harvard Business School Press.
Chilvers, J. (2010). *Sustainable participation? Mapping out and reflecting on the field of public dialogue in science and technology*. Working Paper. Harwell, UK: Sciencewise Expert Resource Centre.
Collingridge, D. (1980). *The social control of technology*. Milton Keynes, UK: Open University Press.
Collins, H. M. (1988). Public experiments and displays of virtuosity: the core-set revisited. *Social Studies of Science*, 18(4), 725–748.
——— (1992). *Changing order: replication and induction in scientific practice*. Chicago: University of Chicago Press.
Collins, H. M., and Pinch, T. (1998). *The golem: what you should know about science*. Cambridge: Cambridge University Press.
Cooper, M. (2007). *Clinical capital – neo-liberalism and the will to experiment (China and the US)*. GBRG Working Paper No. 18. London: Global Biopolitics Research Group, King's College.
David, P. A. (1985). Clio and the economics of QWERTY. *American Economic Review*, 75(2), 332–337.
Davies, G. (2010). Where do experiments end? *Geoforum*, 41(5), 667–670.
Del Vecchio Good, M-J., Good, B. J., Schaffer, C., and Lind, S. E. (1990). American oncology and the discourse on hope. *Culture, Medicine, and Psychiatry*, 14(1), 59–79.
Depledge, M. H., Pleasants, L. J., and Lawton, J. H. (2010). Nanomaterials and the environment: the views of the Royal Commission on Environmental Pollution (UK). *Environmental Toxicology and Chemistry*, 29(1), 1–4.
Dorbeck-Jung, B., and Shelley-Egan, C. (2013). Meta-regulation and nanotechnologies: the challenge of responsibilisation within the European Commission's Code of Conduct for Responsible Nanosciences and Nanotechnologies Research. *NanoEthics*, 7(1), 55–68.
Douglas, H. E. (2003). The moral responsibilities of scientists (tensions between autonomy and responsibility). *American Philosophical Quarterly*, 40(1), 59–68.
——— (2009). *Science, policy, and the value-free ideal*. Pittsburgh: University of Pittsburgh Press.
Duflo, E., Banerjee, A., Glennerster, R., and Kinnan, C. G. (2013). *The miracle of microfinance? Evidence from a randomized evaluation*. No. w18950. Cambridge, MA: National Bureau of Economic Research.
EC (2008). *A code of conduct for responsible nanosciences and nanotechnologies research*. Brussels: European Commission.
Economist, The (2010). We all want to change the world. *The Economist*, 31 March. Available online at http://www.economist.com/node/15814427 (accessed 10 July 2014).
Elias, N. (1956). Problems of involvement and detachment. *British Journal of Sociology*, 7(3), 226–252.
Ellul, J. (1962). The technological order. *Technology and Culture*, 3(4), 394–421.
Evans, R., Kotchetkova, I., and Langer, S. (2009). Just around the corner: rhetorics of progress and promise in genetic research. *Public Understanding of Science*, 18(1), 43–59.
Feenberg, A. (1991). *Questioning technology*. London: Routledge.
Felt, U., and Wynne, B. (2007). *Taking European knowledge society seriously*. Luxembourg: European Commission, DG for Research.

Fisher, E. (2007). *Ethnographic invention: probing the capacity of laboratory decisions.* NanoEthics, 1(2), 155–165.

Fisher, E., Mahajan, R. L., and Mitcham, C. (2006). Midstream modulation of technology: governance from within. *Bulletin of Science, Technology & Society*, 26(6), 485–496.

Fortun, M. (2005). For an ethics of promising, or: a few kind words about James Watson. *New Genetics and Society*, 24(2), 157–174.

Freeman, C. (1987). *Technology and economic performance: lessons from Japan.* London: Pinter.

Fuller, S. (1993). *Philosophy, rhetoric, and the end of knowledge: the coming of science and technology studies.* Madison: University of Wisconsin Press.

——— (2012). Precautionary and proactionary as the new right and the new left of the twenty-first century ideological spectrum. *International Journal of Politics, Culture, and Society*, 25(4), 157–174.

Gardiner, S. M. (2010). Is 'arming the future' with geoengineering really the lesser evil? Some doubts about the ethics of intentionally manipulating the climate system. In S. Gardiner, S. Caney, D. Jamieson and H. Shue (eds), *Climate ethics: essential readings* (pp. 284–312). Oxford: Oxford University Press.

Garrett, J. (2008). New York to Paris the hard way, 100 years ago. *New York Times*, 10 February.

Geels, F. W., and Smit, W. A. (2000). Failed technology futures: pitfalls and lessons from a historical survey. *Futures*, 32(9), 867–885.

Gibbons, M., Limoges, C., Nowotny, H., Schwartzman, S., Scott, P., and Trow, M. (1994). *The new production of knowledge: the dynamics of science and research in contemporary societies.* London: Sage.

Gisler, M., Sornette, D., and Woodard, R. (2011). Innovation as a social bubble: the example of the Human Genome Project. *Research Policy*, 40(10), 1412–1425.

Goodin, R. E., and Dryzek, J. S. (2006). Deliberative impacts: the macro-political uptake of mini-publics. *Politics & Society*, 34(2), 219–244.

Gooding, D., Pinch, T., and Schaffer, S. (1989). Introduction: some uses of experiment. In D. Gooding, T. Pinch, and S. Schaffer (eds), *The uses of experiment: studies in the natural sciences* (pp. 1–27). Cambridge: Cambridge University Press.

Gray, J. (1999). The New Statesman essay – The myth of progress. *New Statesman*, 9 April 1999.

Grinbaum, A., and Groves, C. (2013). What is 'responsible' about responsible innovation? Understanding the ethical issues. In R. Owen, J. Bessant and M. Heintz (eds), *Responsible innovation: managing the responsible emergence of science and innovation in society* (pp. 119–142). Chichester: Wiley.

Gross, M., and Krohn, W. (2005). Society as experiment: sociological foundations for a self-experimental society. *History of the Human Sciences*, 18(2), 63–86.

Groves, C. (2013). Horizons of care: from future imaginaries to responsible research and innovation. In K. Konrad, C. Coenen, A. Dijkstra, C. Milburn and H. van Lente, *Studies of new and emerging technologies*, Vol. 4: *Shaping emerging technologies: governance, innovation, discourse* (pp. 185–202). Amsterdam: IOS Press.

Guston, D. H. (2004). Responsible innovation in the commercialized university. In D. G. Stein (ed.), *Buying in or selling out? The commercialization of the American research university* (pp. 161–174). New Brunswick, NJ: Rutgers University Press.

——— (2008). Innovation policy: not just a jumbo shrimp. *Nature*, 454(7207), 940–941.

——— (2012). The pumpkin or the tiger? Michael Polanyi, Frederick Soddy, and anticipating emerging technologies. *Minerva*, 50(3), 363–379.

——— (2014). Understanding 'anticipatory governance'. *Social Studies of Science*, 44(2), 218–242.

Guston, D. H., and Sarewitz, D. (2002). Real-time technology assessment. *Technology in Society*, 24(1), 93–109.

Hacking, I. (1983). *Representing and intervening: introductory topics in the philosophy of natural science*. Cambridge: Cambridge University Press.

Hackmann, W. D. (1989). Scientific instruments: models of brass and aids to discovery. In D. Gooding, T. Pinch and S. Schaffer (eds), *The uses of experiment: studies in the natural sciences* (pp. 31–65). Cambridge: Cambridge University Press.

Hamilton, C. (2013). *Earthmasters: the dawn of the age of climate engineering*. New Haven, CT: Yale University Press.

Harremoës, P., Gee, D., MacGarvin, M., Stirling, A., Keys, J., *et al.* (2001). *Late lessons from early warnings: the precautionary principle 1896–2000*. Environmental Issue Report 22. Copenhagen: European Environment Agency.

Haynes, L., Servie, O., Goldacre, B., and Torgerson, D. (2012). *Test, learn, adapt: developing public policy with randomised controlled trials*. London: Cabinet Office, Behavioural Insights Team.

Hedgecoe, A. (2010). Bioethics and the reinforcement of socio-technical expectations. *Social Studies of Science*, 40(2), 163–186.

Horlick-Jones, T., Walls, J., Rowe, G., Pidgeon, N., Poortinga, W., *et al.* (2007). *The GM debate: risk, politics and public engagement*. London: Routledge.

Hughes, T. P. (1993). *Networks of power: electrification in Western society, 1880–1930*. Baltimore: Johns Hopkins University Press.

——— (2004). Afterword. In L. Rosner (ed.), *The technological fix: how people use technology to create and solve problems* (pp. 241–243). New York: Routledge.

Illies, C., and Meijers, A. (2009). Artefacts without agency. *The Monist*, 92(3), 420–440.

Jasanoff, S. (2003). Technologies of humility: citizen participation in governing science. *Minerva*, 41(3), 223–244.

Jasanoff, S., and Kim, S. H. (2009). Containing the atom: sociotechnical imaginaries and nuclear power in the United States and South Korea. *Minerva*, 47(2), 119–146.

Jaud, J-P. (dir.) (2012). *Tous cobayes? All of us guinea pigs now?* DVD video. Mareil-sur-Mauldre, France: J+B Séquences.

John, P. (2013). Experimentation, behaviour change and public policy. *Political Quarterly*, 84(2), 238–246.

Jonas, H. (1985). *The imperative of responsibility: in search of an ethics for the technological age*. Chicago: University of Chicago Press.

Jones, R. (2008). The economy of promises. *Nature Nanotechnology*, 3(2), 65–66.

——— (2009). Are you a responsible nanoscientist? *Nature Nanotechnology*, 4(6), 336.

——— (2012). Geek power? *Soft Machines* blog, 5 September 2012. Available online at http://www.softmachines.org/wordpress/?p=1228 (accessed 26 June 2014).

Joy, B. (2000). Why the future doesn't need us. *Wired*, 8(04), 1–11. Available online at http://www.wired.com/wired/archive/8.04/joy.html (accessed 7 July 2014).

Kearnes, M., and Rip, A. (2009). The emerging governance landscape of nanotechnology. In S. Gammel, A. Lösch and A. Nordmann (eds), *Jenseits von Regulierung: Zum politischen Umgang mit der Nanotechnologie* (pp. 97–121). Berlin: Akademische Verlagsgesellschaft.

Kelly, K. (2010). *What technology wants*. New York: Penguin.

Kintisch, E. (2010). 'Asilomar 2' takes small steps toward rules for geoengineering. *Science*, 328(5974), 22–23.

Kline, R., and Pinch, T. (1996). Users as agents of technological change: the social construction of the automobile in the rural United States. *Technology and Culture*, 37(4), 763–795.

Kramer, A. D. I., Guillory, J. E., and Hancock, J. T. (2014). Experimental evidence of massive-scale emotional contagion through social networks. *Proceedings of the National Academy of Sciences*, 111(24), 8788–8790.

Krohn, W., and Weingart, P. (1987). Commentary: Nuclear power as a social experiment – European political 'fall out' from the Chernobyl meltdown. *Science, Technology, & Human Values*, 12(2), 52–58.

Latour, B. (writing as Jim Johnson) (1988). Mixing humans and nonhumans together: the sociology of a door-closer. *Social Problems*, 298–310.

Latour, B. (1992). Where are the missing masses? The sociology of a few mundane artifacts. In W. E. Bijker and J. Law (eds), *Shaping technology/building society: studies in sociotechnical change* (pp. 225–258). Cambridge, MA: MIT Press.

——— (1998). From the world of science to the world of research? *Science*, 280(5361), 208–209.

——— (1999). *Pandora's hope: essays on the reality of science studies*. Cambridge, MA: Harvard University Press.

——— (2008). 'It's development, stupid!' or, how to modernize modernization. In J. Proctor (ed.), *Postenvironmentalism* (pp. 1–13). Cambridge, MA: MIT Press.

——— (2010). An attempt at a 'compositionist manifesto'. *New Literary History*, 41(3), 471–490.

——— (2011). Love your monsters. In M. Shellenberger and T. Nordhaus (eds), *Love your monsters: postenvironmentalism and the Anthropocene* (pp. 17–23). Oakland, CA: Breakthrough Institute.

Latour, B., and Venn, C. (2002). Morality and technology: the end of the means. *Theory, Culture & Society*, 19(5/6), 247–260.

Latour, B., and Woolgar, S. (1979). *Laboratory life: the social construction of scientific facts*. Beverly Hills: Sage.

Laurent, B. (2011). Technologies of democracy: experiments and demonstrations. *Science and Engineering Ethics*, 17(4), 649–666.

Laurie, G. (2011). Reflexive governance in biobanking: on the value of policy led approaches and the need to recognise the limits of law. *Human Genetics*, 130(3), 347–356.

Levidow, L., and Carr, S. (2007). GM crops on trial: technological development as a real-world experiment, *Futures*, 39(4), 408–431.

Levy, N. (2007). *Neuroethics: challenges for the 21st century*. Cambridge: Cambridge University Press.

Lezaun, J., and Soneryd, L. (2007). Consulting citizens: technologies of elicitation and the mobility of publics. *Public Understanding of Science*, 16(3), 279–297.

Liao, S. M., Sandberg, A., and Roache, R. (2012). Human engineering and climate change. *Ethics, Policy & Environment*, 15(2), 206–221.

Lüthje, C., Herstatt, C., and von Hippel, E. (2005). User-innovators and 'local' information: the case of mountain biking. *Research Policy*, 34(6), 951–965.

Lynch, M. (2000). Against reflexivity as an academic virtue and source of privileged knowledge. *Theory, Culture & Society*, 17(3), 26–54.

Macnaghten, P., and Chilvers, J. (2012). Governing risky technologies. In S. Lane, F. Klauser and M. B. Kearnes (eds), *Critical risk research: practices, politics and ethics* (pp. 99–124). Chichester: Wiley-Blackwell.

Manzi, J. (2012). *Uncontrolled: the surprising payoff of trial-and-error for business, politics, and society*. New York: Basic Books.

Marres, N. (2012). The experiment in living. In C. Lury and N. Wakeford (eds), *Inventive methods: the happening of the social* (pp. 76–95). London: Routledge.

Marris, C., and Rose, N. (2010). Open engagement: exploring public participation in the biosciences. *PLoS Biology*, 8(11), e1000549.

——— (2012). Let's get real on synthetic biology. *New Scientist*, 214(2868), 28–29.

Merton, R. K. (1973). The normative structure of science. Reprinted in *The sociology of science: theoretical and empirical investigations* (pp. 223–280). Chicago: University of Chicago Press.

Michael, M. (2000). Futures of the present: from performativity to prehension. In N. Brown, B. Rappert and A. Webster (eds), *Contested futures: a sociology of prospective techno-science* (pp. 21–39). Aldershot: Ashgate.

Miles, J. (2011). The billionaire king of techtopia. *Details*, 1 September 2011. Available online at http://www.details.com/culture-trends/critical-eye/201109/peter-thiel-billionaire-paypal-facebook-internet-success?currentPage=2 (accessed 2 July 2014).

Miller, C. A., and Edwards, P. N. (eds) (2001). *Changing the atmosphere: expert knowledge and environmental governance*. Cambridge, MA: MIT Press.

Mitcham, C. (2003). Co-responsibility for research integrity. *Science and Engineering Ethics*, 9(2), 273–290.

Mol, A. (2008). *The logic of care: health and the problem of patient choice*. Abingdon: Routledge.

Morozov, E. (2013). *To save everything, click here: technology, solutionism, and the urge to fix problems that don't exist*. London: Penguin UK.

NCPHSBBR (1979). *Ethical principles and guidelines for the protection of human subjects of research*. Belmont report. Washington, DC: National Commission on the Protection of Human Subjects of Biomedical and Behavioral Research, US Department of Health, Education and Welfare.

Nelson, R. R. (2011). *The moon and the ghetto* revisited. *Science and Public Policy*, 38(9), 681–690.

Nerlich, B., Dingwall, R., and Clarke, D. D. (2002). The book of life: how the completion of the Human Genome Project was revealed to the public. *Health*, 6(4), 445–469.

Newby, H. (1992). One society, one Wissenschaft: a 21st century vision. *Science and Public Policy*, 19(1), 7–14.

Nordmann, A. (2007). If and then: a critique of speculative nanoethics. *NanoEthics*, 1(1), 31–46.

——— (2010). A forensics of wishing: technology assessment in the age of technoscience. *Poiesis & Praxis*, 7(1/2), 5–15.

Nowotny, H. (2005). The changing nature of public science. In H. Nowotny, D. Pestre, E. Schmidt-Aßmann, H. Schulze-Fielitz and H-H. Trute, *The public nature of science under assault: politics, markets, science and the law* (pp. 1–27). Berlin and Heidelberg: Springer.

Nuremberg Military Tribunals (1949). Nuremberg code. In *Trials of war criminals before the Nuremberg Military Tribunals under Control Council Law No. 10*, Vol. 2 (pp. 181–182). Washington, DC: U.S. Government Printing Office.

Owens, S. (1995). From 'predict and provide' to 'predict and prevent'? Pricing and planning in transport policy. *Transport Policy*, 2(1), 43–49.

Park, R. E. (1929). The city as a social laboratory. In T. V. Smith and L. D. White (eds), *Chicago: an experiment in social science research* (pp. 1–19). University of Chicago Social Science Studies Series, No. 17. Chicago: University of Chicago Press.

Pellizzoni, L. (2004). Responsibility and environmental governance. *Environmental Politics*, 13(3), 541–565.
Perrow, C. (1984). *Normal accidents: living with high-risk technologies.* New York: Basic Books.
——— (2010). The meltdown was not an accident. In M. Lounsbury and P. M. Hirsch (eds), *Research in the Sociology of Organizations*, Vol. 30: *Markets on trial: the economic sociology of the U.S. financial crisis: part A* (pp. 309–330). Bingley, UK: Emerald Group.
——— (2011). Fukushima and the inevitability of accidents. *Bulletin of the Atomic Scientists*, 67(6), 44–52.
Pfotenhauer, S. M., Jones, C. F., Saha, K., and Jasanoff, S. (2012). Learning from Fukushima. *Issues in Science and Technology*, 28(3), 79–84.
Pickstone, J. V. (2001). *Ways of knowing: a new history of science, technology, and medicine.* Chicago: University of Chicago Press.
Polanyi, M. (1962). The republic of science: its political and economic theory. *Minerva*, 1(1), 54–73.
Prigogine, I. (1997). *The end of certainty: time, chaos, and the new laws of Nature.* New York: Simon & Schuster.
Rabinow, P. (2009). *Marking time: on the anthropology of the contemporary.* Princeton: Princeton University Press.
Rabinow, P., and Bennett, G. (2012). *Designing human practices: an experiment with synthetic biology.* Chicago: University of Chicago Press.
Radder, H. (2009). The philosophy of scientific experimentation: a review. *Automated Experimentation*, 1(1), 2.
Ravetz, J. R. (1975). . . . et augebitur scientia. In R. Harré (ed.), *Problems of scientific revolution: progress and obstacles to progress in the sciences* (pp. 42–75). London: Oxford University Press.
Rayner, S. (2004). The novelty trap: why does institutional learning about new technologies seem so difficult? *Industry & Higher Education*, 18(6), 349–355.
——— (2012). Uncomfortable knowledge: the social construction of ignorance in science and environmental policy discourses. *Economy and Society*, 41(1), 107–125.
Rayner, S., and Cantor, R. (1987). How fair is safe enough? The cultural approach to societal technology choice. *Risk Analysis*, 7(1), 3–9.
Renn, O., Webler, T., and Wiedemann, P. (1995). *Fairness and competence in citizen participation: evaluating models for environmental discourse.* Dordrecht: Kluwer.
Rheinberger, H-J. (1997). *Toward a history of epistemic things: synthesizing proteins in the test tube.* Stanford, CA: Stanford University Press.
Richards, N. M., and King, J. H. (2013). Three paradoxes of big data. *Stanford Law Review Online*, 66(41), 41–46.
Rip, A. (1994). Science & technology studies and constructive technology assessment. Keynote speech to EASST Conference, Budapest, 28–31 August 1994. *EASST Newsletter*, 13(3), 11–16.
Rip, A., Misa, T. J., and Schot, J. (eds). (1995). *Managing technology in society.* London and New York: Pinter.
Rose, N. (1999). *Powers of freedom: reframing political thought.* Cambridge: Cambridge University Press.
Rosenblueth, A., and Wiener, N. (1945). The role of models in science. *Philosophy of Science*, 12(4), 316–321.
Rothstein, H. (2007). Talking shop or talking turkey? Institutionalizing consumer representation in risk regulation. *Science, Technology, & Human Values*, 32(5), 582–607.

Royal Society (2006). *Survey of factors affecting science communication by scientists and engineers*. London: Royal Society. Available online at https://royalsociety.org/~/media/Royal_Society_Content/policy/publications/2006/1111111395.pdf (accessed 19 June 2014).

——— (2009). *Reaping the benefits: science and the sustainable intensification of global agriculture*. London: Royal Society.

RS–RAEng (2004). *Nanoscience and nanotechnologies: opportunities and uncertainties*. London: Royal Society & Royal Academy of Engineering.

Sandel, M. (2004). The case against perfection. *Atlantic Monthly*, 293(3), 51–62.

Sarewitz, D., and Nelson, R. (2008). Three rules for technological fixes. *Nature*, 456(7224), 871–872.

Schaffer, S. (1989). Glass works: Newton's prisms and the uses of experiment. In D. Gooding, T. Pinch and S. Schaffer (eds), *The uses of experiment: studies in the natural sciences* (pp. 67–104). Cambridge: Cambridge University Press.

Schneider, J. W. (1985). Social problems theory: the constructionist view. *Annual Review of Sociology*, 11, 209–229.

Schuurbiers, D. (2011). What happens in the lab: applying midstream modulation to enhance critical reflection in the laboratory. *Science and Engineering Ethics*, 17(4), 769–788.

Sciencewise (2010). *Science, governance and public engagement*. TNS-BMRB Report JN 219638 for the Department of Business, Innovation and Skills. Available online at http://www.sciencewise-erc.org.uk/cms/science-trust-and-public-engagement-2/ (accessed 19 June 2014).

Shamir, R. (2008). The age of responsibilization: on market-embedded morality. *Economy and Society*, 37(1), 1–19.

Shapin, S. (1988a). The house of experiment in seventeenth-century England. *Isis*, 79(3), 373–404.

——— (1988b). Following scientists around. *Social Studies of Science*, 18(3), 533–550.

Shapin, S., and Schaffer, S. (1985). *Leviathan and the air-pump*. Princeton: Princeton University Press.

Shapira, P., Youtie, J., and Porter, A. L. (2010). The emergence of social science research on nanotechnology. *Scientometrics*, 85(2), 595–611.

Simakova, E., and Coenen, C. (2013). Visions, hype, and expectations: a place for responsibility. In R. Owen, J. Bessant and M. Heintz (eds), *Responsible innovation: managing the responsible emergence of science and innovation in society* (pp. 241–266). Chichester: Wiley.

Sismondo, S. (2008). Science and technology studies and an engaged program. In E. J. Hackett, O. Amsterdamska, M. Lynch and J. Wajcman, *The handbook of science and technology studies*, 3rd edn (pp. 13–31), Cambridge, MA: MIT Press.

Stahl, B. C. (2013). Responsible research and innovation: the role of privacy in an emerging framework. *Science and Public Policy*, 40(6), 708–716.

Stengers, I. (1997). *Power and invention*. Minneapolis: University of Minnesota Press.

Stilgoe, J. (2007). *Nanodialogues: experiments in public engagement with science*. London: Demos.

——— (2011). A question of intent. *Nature Climate Change*, 1(7), 325–326.

——— (2013). Foreword: Why responsible innovation? In R. Owen, J. Bessant and M. Heintz (eds), *Responsible innovation: managing the responsible emergence of science and innovation in society* (pp. xi–xv). Chichester: Wiley.

Stilgoe, J., Irwin, A., and Jones, K. (2006). *The received wisdom: opening up expert advice*. London: Demos.

Stilgoe, J., Lock, S. J., and Wilsdon, J. (2014). Why should we promote public engagement with science? *Public Understanding of Science*, 23(1), 4–15.
Stilgoe, J., Owen, R., and Macnaghten, P. (2013). Developing a framework for responsible innovation. *Research Policy*, 42(9), 1568–1580.
Stirling, A. (2007). A general framework for analysing diversity in science, technology and society. *Journal of the Royal Society Interface*, 4(15), 707–719.
——— (2008). 'Opening up' and 'closing down': power, participation, and pluralism in the social appraisal of technology. *Science, Technology, & Human Values*, 33(2), 262–294.
——— (2010). Keep it complex. *Nature*, 468(7327), 1029–1031.
Strydom, P. (1999). The challenge of responsibility for sociology. *Current Sociology*, 47(3), 65–82.
Sudjic, D. (2009). *The language of things. Design, luxury, fashion, art: how we are seduced by the objects around us.* London: Penguin UK.
Svedin, U. (2009). *New worlds – new solutions: research and innovation as a basis for developing Europe in a global context.* Final Report. New Worlds – New Solutions Conference, Lund, Sweden, 7–8 July 2009. Online available at http://www.vr.se/download/18.29b9c5ae1268d01cd5c8000631/New_Worlds_New_Solutions_Report.pdf (accessed 16 February 2012).
Szerszynski, B., Kearnes, M., Macnaghten, P., Owen, R., and Stilgoe, J. (2013). Why solar radiation management geoengineering and democracy won't mix. *Environment and Planning A*, 45(12), 2809–2816.
Thomke, S., von Hippel, E., and Franke, R. (1998). Modes of experimentation: an innovation process – and competitive – variable. *Research Policy*, 27(3), 315–332.
Tollefson, J. (2010). Geoengineers get the fear. *Nature*, 464(7289), 656.
Urry, J. (2007). *Mobilities*. Cambridge: Polity Press.
——— (1999). *Automobility, car culture and weightless travel: a discussion paper*. Lancaster: Lancaster University. http://www.lancs.ac.uk/fass/sociology/research/publications/papers/urry-automobility.pdf.
van den Hoven, M. J. (1998). Moral responsibility, public office and information technology. In I. T. M. Snellen and W. B. H. J. van de Donk (eds), *Public administration in an information age: a handbook* (pp. 97–112). Amsterdam: IOS Press.
van de Poel, I. (2011). Nuclear energy as social experiment. *Ethics, Policy & Environment*, 14(3), 285–290.
van Lente, H., and Rip, A. (1998). The rise of membrane technology: from rhetorics to social reality. *Social Studies of Science*, 28(2), 221–254.
Verbeek, P-P. (2006). Materializing morality: design ethics and technological mediation. *Science, Technology, & Human Values*, 31(3), 361–380.
Volti, R. (1995). *Society and technological change*, 3rd edn. New York: St. Martin's Press.
von Hippel, E. (2005). *Democratizing innovation*. Cambridge, MA, and London: MIT Press.
von Schomberg, R. (2007). *From the ethics of technology towards an ethics of knowledge policy & knowledge assessment*. Working Document. Brussels: European Commission, DG for Research. Available online at http://ec.europa.eu/research/science-society/pdf/ethicsofknowledgepolicy_en.pdf (accessed 19 June 2014).
Weinberg, A. M. (1966). Can technology replace social engineering? *Bulletin of the Atomic Scientists*, 22(10), 4–8.
——— (1972). Science and trans-science. *Minerva*, 10(2), 209–222.
Weinstein, M. (2001). A public culture for guinea pigs: US human research subjects after the Tuskegee study. *Science as Culture*, 10(2), 195–223.

Whitelegg, J. (1997). *Critical mass: transport environment and society in the twenty-first century*. London: Pluto Press.

Wildavsky, A. B. (1988). *Searching for safety*. New Brunswick, NJ: Transaction.

Wilsdon, J., and Willis, R. (2004). *See-through science: why public engagement needs to move upstream*. London: Demos.

Winner, L. (1977). *Autonomous technology: technics-out-of-control as a theme in political thought*. Cambridge, MA: MIT Press.

——— (1983). Technologies as forms of life. In R. S. Cohen and M. W. Wartofsky, *Epistemology, methodology, and the social sciences* (pp. 249–263). Dordrecht: Kluwer.

——— (1993). Upon opening the black box and finding it empty: social constructivism and the philosophy of technology. *Science, Technology, & Human Values*, 18(3), 362–378.

Wynne, B. (1988). Unruly technology: practical rules, impractical discourses and public understanding. *Social Studies of Science*, 18(1), 147–167.

——— (1992). Uncertainty and environmental learning: reconceiving science and policy in the preventive paradigm. *Global Environmental Change*, 2(2), 111–127.

——— (1993). Public uptake of science: a case for institutional reflexivity. *Public Understanding of Science*, 2(4), 321–337.

——— (2005). Risk as globalizing 'democratic' discourse? Framing subjects and citizens. In M. Leach, I. Scoones and B. Wynne (eds), *Science and citizens: globalization and the challenge of engagement* (pp. 66–82). London: Zed Books.

——— (2006). Afterword. In M. Kearnes, P. Macnaghten and J. Wilsdon, *Governing at the nanoscale: people, policies and emerging technologies* (pp. 70–78). London: Demos.

——— (2011a). Lab work goes social, and vice versa: strategising public engagement processes. Commentary on 'What happens in the lab does not stay in the lab: applying midstream modulation to enhance critical reflection in the laboratory'. *Science and Engineering Ethics*, 17(4), 791–800.

——— (2011b). *Rationality and ritual: participation and exclusion in nuclear decision-making*, 2nd edn. London: Earthscan.

Yarow, J. (2013). Google CEO Larry Page wants a totally separate world where tech companies can conduct experiments on people. *Business Insider*, 16 May. Available online at http://www.businessinsider.com/google-ceo-larry-page-wants-a-place-for-experiments-2013-5#ixzz36gCMZYJo (accessed 2 July 2014).

Zimmer, C. (2013). Bringing them back to life. *National Geographic*, 2013(4).

Zumbach, L. (2013). Software engineers need a crash course in ethics. *Slate*, 9 September. Available online at http://mobile.slate.com/blogs/future_tense/2013/09/09/software_engineers_need_a_crash_course_in_ethics.html (accessed 10 July 2014).

3 Rethinking the unthinkable

As discussed in the previous chapters, geoengineering is an idea rather than a technology. It is impossible to write a definitive history of an idea, particularly an unattractive idea whose spread has been limited by a degree of scientific self-censorship. The histories that geoengineering is already adopting for itself are tied to particular futures, emphasising aspects of continuity and disconnection to draw particular lessons. There are two major versions of geoengineering history. The first is the story adopted by many researchers working on geoengineering. It is a story of a dangerous idea that was for many decades shrouded in a taboo. Geoengineering was seen by many scientists as off-limits because of its association with Cold War technological utopianism and a widespread political commitment among climate scientists at the end of the twentieth century towards mitigation of greenhouse gas emissions as the only viable approach to climate change. With the demonstrable failure of this policy approach, the story goes, the taboo became unsustainable and was eventually lifted by the publication in 2006 of a paper by Paul Crutzen. The contribution of this Nobel laureate is seen as moving geoengineering from the fringes into the mainstream of science and policy discussion. This history is one of disjuncture.

The second story we can tell about the rise of geoengineering is more problematic and invites a different debate about the responsibilities of scientists. This story has no sharp disconnect between climate science and geoengineering. It sees ideas about geoengineering running throughout the post-war history of climate science, and it ties the recent rehabilitation of the idea to the scientism of hybrid constructions such as 'planetary boundaries' and the 'Anthropocene'. From this perspective, the idea of engineering the climate is merely the next step along a road towards greater prediction in climate science and greater control of the 'experiment' of climate change.

The history of disjuncture forms part of the current rhetoric of geoengineering research. It has been written and adopted by the scientists who have been most involved in the recent renewal of enthusiasm. David Keith tells how 'for decades a de-facto taboo against serious work on geoengineering discouraged quantitative work; little was done. Paul Crutzen's 2006 paper arguing for geoengineering research broke that taboo' (Keith 2013, p. 92). The story is one of revelation. A set of technical possibilities was announced to the world by a scientist, Paul

Crutzen, who could not be ignored. As we will see later in this chapter, 2006 does indeed seem to mark a turning point in geoengineering research. But the emphasis on this moment – a sudden awakening of scientific realisation that climate control was possible – as the starting point for the current phase of geoengineering research obscures a more complicated history.

The major historical work to draw longer lines of continuity, James Fleming's (2010) *Fixing the Sky*, tells a story of geoengineering as a scaling-up of the ambitions of the 'rainmakers', early weather modification salesmen whose ideas had little scientific basis and little hope of success. Fleming overstates the scientific links between weather and climate modification, but his message is that geoengineering repeats many of the patterns of hubris shown around past technologies and risks repeating those mistakes if the moral and practical lessons of history are ignored.

Fleming's book follows an earlier attempt to historically situate geoengineering, by David Keith (2000). Although Keith in later talks and writings leans more towards the story of disconnect, his 2000 piece provides a history as both observer of and participant in what was then a very small world. He draws lines of continuity with weather modification schemes in the USA and the USSR, speculation about terraforming other planets, and contemporary climate policy. Keith has become the most prominent spokesperson for geoengineering research, in part because of his length of experience – he claims to have produced the first proper assessment of geoengineering proposals in 1992 (Keith and Dowlatabadi 1992) – but he also claims broad coverage of different areas of expertise, including science, engineering and public policy.

The idea of taking control of and taking responsibility for the climate is not as clear-cut as the history of geoengineering-as-taboo would suggest, nor is its heritage just in the aberrant ambitions of the rainmakers who would seek to control local weather. In explaining the emergence of geoengineering, we should also look at more conventional twentieth-century science. The growth of Earth system science and its connections with policies and practices ranging from weather forecasting and carbon markets to insurance and biodiversity offsetting make it easy for scientists to argue that humans are already, in some respects, responsible for the climate. With this realisation, the step to geoengineering does not seem so vast. The histories of geoengineering and conventional climate science are more entwined than many climate scientists would admit. Geoengineering has been made 'thinkable' not just by the removal of an informal taboo but also by decades of environmental science that had increased the confidence of scientists in understanding the relationship between humans and the climate. This has important implications for scientific responsibility. Understanding this more complicated history puts climate science and climate models back into the emerging technology of geoengineering, rather than detaching them.

If we are considering the responsibilities of scientists within and around geoengineering, we should ask how geoengineering, previously considered beyond the pale, started to become normal. How have the technologies under consideration become stabilised, even though they are still imaginary? How have particular questions of governance and ethics emerged as important? If we are interested in exploring a democratic approach to geoengineering, we should take an interest in how geoengineering has become, according to Dilling and Hauser, 'a highly technical issue' that 'discourages entry and dialogue by non-specialists' (Dilling and Hauser 2013, p. 556). As geoengineering has become a research agenda, certain questions have emerged as hugely important, and others have been shut out.

In this chapter, I look at some of the stories told about the origins and development of geoengineering to examine some assumptions and tensions within contemporary geoengineering research. I look at some important developments in the interplay of climate science and politics in the late twentieth century and consider how motivations of scientific understanding have become entangled in those of scientific control. I conclude by asking whether it would be possible or desirable to unthink the idea of geoengineering.

Knowing the climate; making the climate

Contemporary life sciences blur the distinction between understanding and intervening (Nowotny and Testa 2010). Biotechnology and the biosciences are now intertwined such that, especially with the turn towards synthetic biology, 'to know life, is to remake life' (Nowotny 2007). With respect to the climate and geoengineering, scientists have taken pains to separate knowledge from control, but this distinction is a relatively recent one. A look back at the early history of climate science reveals greater ambition.

The idea of planetary climate control has captured the scientific imagination since before weather systems were well understood. As Zach Horton (2014) has described, Francis Bacon imagined weather control being an important part of the activities of 'Salomon's House' – his utopian scientific institution. J. D. Bernal, who was, among his other talents, a founding father of science policy scholarship, wrote in his 1939 book *The Social Function of Science* that

> it will no longer be a question of adapting man to the world but the world to man. For instance, the present Arctic with its wastes of tundra, glacier, and sea ice is a legacy of the geological accident of the Ice Age. It will disappear in time, leaving the world a much pleasanter place, but there is no reason why man should not hasten the process. By an intelligent diversion of warm ocean-currents together with some means of colouring snow so that the sun could melt it, it might be possible to keep the Arctic ice-free for one summer, and that one year might tip the balance and permanently change the climate of the northern hemisphere.
>
> (Bernal 1939 [2010 edn], pp. 379–380)[1]

In the immediate aftermath of World War II, scientists sought to modernise meteorology (Harper 2008). Central to this project was John von Neumann, who wanted a mathematical theory of general circulation that would, enabled by computers, allow for prediction and then control. Kristine Harper quotes von

Neumann arguing for 'the first steps toward influencing the weather by rational, human intervention . . . since the effects of any hypothetical intervention will have become calculable' (von Neumann, quoted in Harper 2008, pp. 4–5). While few meteorologists would have publicly endorsed von Neumann's aim, Harper describes how their science was driven forward not just by fascination with the complexity of the environments but also by urgent operational requirements from weather forecasters, fuelled by the military, the transport industry, insurers and a growing public appetite.

The ideas of climate control in circulation during this time were extensions of previous weather-modification proposals (Fleming 2010). The idea was to find a lever that would allow for what von Neumann called 'jiggling' of the planet. In the 1960s the term 'geoengineering' started to be used in relation to the climate, rather than just conventional land-based engineering. Lloyd Berkner, one of the driving forces behind the 1957–8 International Geophysical Year, wrote in 1962 about what he called 'applied geoscience, which might properly be called geoengineering':

> The explosive development of geoscience in the past decade bids us look for an equally explosive advance in geoengineering. Man is becoming aware of the controls and adjustments that can be exercised with respect to the environment in which he must exist. Out of our new geoscience will emerge a most extensive series of engineering applications of very general importance.
> (Berkner 1962, p. 2182)[2]

For Berkner, these engineering applications relate to weather forecasting, radio waves, ballistics and sonar, but he leaves open the possibility of various others relating to aspects of the Earth (in addition to other planets and moons), its environment and space. It was not until the 1970s that 'geoengineering' came to acquire something like its current usage (see Marchetti 1977).

In 1966, a study from the US National Science Foundation on weather and climate modification described the fundamental asymmetry of mankind's contest against the weather. Estimating the amount of kinetic energy involved in various atmospheric subsystems, from tornadoes and thunderstorms up to the northern hemisphere's general climatic circulation, the Special Commission on Weather Modification concluded that it makes little sense to attempt to play the weather at its own game (NSF 1966). So the commission turned its attention to the instabilities inherent in weather and climate systems that might suggest 'triggering' opportunities for changing the weather and climate. Importantly, the commission used the term 'modification' to include both inadvertent changes to the atmosphere through pollution and deliberate intervention.

This report's gestation overlapped with Edward Lorenz's development in the 1960s of chaos theory, popularised as the 'butterfly effect'. Lorenz's work would reveal the extent to which weather could be altered by tiny changes, and it would subsequently be used to justify the ambitions of weather modifiers such as Ross Hoffman, who advocated 'steering hurricanes' away from centres of population

(Fleming 2010, p. 197). Lorenz's work came too late to influence the commission (indeed, the only mention of 'chaos' in the report relates to the sociopolitical implications of granting property rights over water in the atmosphere [NSF 1966, p. 103]). But the idea of taking advantage of the instability of weather was central to the report's conclusions. Less than a year before, Lyndon Johnson's President's Science Advisory Committee, in its report *Restoring the Quality of Our Environment*, had diagnosed the carbon dioxide problem and suggested climate modification as the only sensible response (PSAC 1965).

Joseph Fletcher from the RAND think tank published a report in 1968 on climate change that included an argument for research into, and subsequent control of, climate. Fletcher assumed that growing scientific understanding of the climate would inevitably 'trigger an avalanche of "climatic experiments" testing the predictions of the improved theory of climate' (quoted in Fleming 2010, p. 239). Fletcher developed the idea of climate 'triggers' from the National Science Foundation (1966) report. The strategy he presents is linear:

> It is convenient to think of progress toward climate control in four stages – observation, understanding, prediction, and control. We must observe how nature behaves before we can understand why, we must understand before we can predict, and we must be able to predict the outcome before we undertake measures for control.
>
> (Quoted in Fleming 2010, p. 240)

Fletcher went on to bemoan the ability of the climate models of the time to perform this task of prediction, but argued that they would be up to the job by 1973 (Fleming 2010, p. 240).

Contemporary geoengineering researchers might respond that Fletcher's and von Neumann's ambitions were quashed by some important realisations, namely that climate science soon revealed the climate to be less predictable than assumed and that American military-industrial techno-optimism could not survive the end of the Cold War. But the ambitions of this generation of cold warriors continued through people like Edward Teller and Lowell Wood. Teller, the 'father of the hydrogen bomb', was a Hungarian émigré who had joined the Manhattan Project before setting up the Lawrence Livermore National Laboratory after World War II. Lowell Wood, also a physicist, joined Teller at Lawrence Livermore in the 1960s and 1970s. They worked together on military projects, including the Strategic Defense Initiative ('Star Wars') and geoengineering projects.

The atomic bombs dropped at the end of World War II were discussed and developed in secret. The hydrogen bombs that followed were more openly debated among scientists (Galison and Bernstein 1989). Technical and moral arguments divided scientists. Those in favour of the bomb, such as Teller (Steven Shapin [2002] notes that Teller 'never saw a new type of nuclear weapon that he didn't like'), broadened their arguments to sell the technology to their sceptics. The power of hydrogen bombs was pointed to peaceful ends in Eisenhower's Project Plowshare. Teller, for his part, proposed schemes such as Project Chariot, in which the coastline of Alaska would be granted a new harbour through the pinpointed explosion of up to six enormous bombs.

Alongside research into martial weather modification (Fleming 2010), growing scientific knowledge at a planetary scale prompted the development of 'synoptic' weapons (Hamblin 2013). In the late 1950s and early 1960s, the USA and the USSR both attempted to understand the Van Allen belt surrounding the Earth by disturbing it with high-altitude hydrogen bomb explosions. The US effort, Project Argus, was dubbed the 'greatest experiment of all time' by the scientists involved (Hamblin 2013, p. 122). Prefiguring later scientific concerns about geoengineering, astronomer Bernard Lovell responded to Project Argus that 'no government has the right to change the environment in any significant way without prior international study and agreement' (quoted in Fleming 2012). This blend of experimentation and aggression was curtailed by the 1963 test ban treaty and by the Environmental Modification Convention (ENMOD) in 1978.

The weapons researchers of the Cold War sought new uses for their tools to meet the new geopolitics. Research on weather modification during the Vietnam War, which prompted the ENMOD treaty, was redeployed against the global target of climate change. What were previously explicit intentions – to change local weather – became mere side effects, for scientific assessment and mitigation, of the grander plan. In some cases, unintended consequences such as effects on crops were sold as positive externalities. Around the turn of the millennium, Teller, Wood and their colleague Roderick Hyde published a pair of reports analysing the prospects for solar radiation management (SRM), concluding that it would pose economic benefits and minimal risk (Teller *et al.* 1997, 2002). By 2002, as Hamilton (2013, p. 128) narrates, the technological hubris of these authors is quite breathtaking.

As I will describe, the rehabilitation of geoengineering research required scientists to create some distance between themselves and these Cold War characters. Hamilton (2013) argues that Cold War culture still has a strong influence on geoengineering research. But we should recognise, however, that the characters involved are, in the main, very different. Much recent geoengineering research emanates from scientists who had associations with nuclear disarmament and peace movements. Echoes of macho hubris still enter geoengineering discussions. Early geoengineering speculation was justified in much the same way as a lot of Cold War research – the enemy may be thinking of it, so we should too. Now the military posturing is gone, but geoengineering research still has a blend of 'offensive' and 'defensive' justifications (see Dando *et al.* 2006 in the case of biological weapons). However, the story of geoengineering as a Cold War hangover is partial at best. Understanding the emergence of geoengineering requires closer engagement with mainstream science and science policy.

Climate science and climate politics

In December 2009, negotiators from the countries of the United Nations met in Copenhagen to tackle climate change. According to the plan agreed in Bali in 2007, a new framework for a concerted global approach to the mitigation of climate change, to start in 2012, would be agreed at the meeting, but few were optimistic. A few weeks before, email and documents hacked from a server at the University of East Anglia's Climate Research Unit had been circulated among climate change sceptics. These had been offered as evidence of a scientific cover-up. The episode, which came to be known as Climategate, added to the fevered atmosphere among scientists and environmentalists with long experience of the climate debate.

There were hopes that the Copenhagen meeting would produce a legally binding contract, with a substantial budget attached, to replace the Kyoto Protocol. Instead, we got the Copenhagen 'Accord', a wispy agreement. This was diplomatically sold to the world by UN Secretary-General Ban Ki-moon as an 'essential beginning', but those close to the process recognised that it marked the end of a disastrous set of negotiations.

These negotiations took place at a difficult time, in the middle of a long and deep global recession during which many countries quietly moved environmental sustainability down their lists of priorities as they concentrated on economic recovery and growth. Ironically, this recession achieved what no policy measures had managed over the previous three decades: a reduction in the rate of carbon dioxide emissions. For exasperated scientists and environmentalists, Copenhagen provided further cause for despair, if not the final straw. Martin Rees, then president of the Royal Society, captured the mood, saying 'As a global community, we now move one step closer to a humanitarian crisis, where those least able to adapt will be worst affected' (quoted in ENS 2009). The sort of fatalism that characterises our attitude to the weather (particularly in Britain) descended over climate scientists as they observed the inadequacies of international policymaking.

Given the complexity (or 'wickedness'; Rayner 2012) of global climate change as a policy issue, a cynical political analyst might express surprise that it has risen up political agendas at all. Climate change became an important feature of US and European politics in the 1980s, due in part to the political entrepreneurship of some prominent climate scientists. Governments and scientific institutions had until this point been relatively relaxed about, or merely interested in, the changing climate (Edwards 2001). The policy responses that followed were the result of a particular framing of the issue (Sarewitz and Pielke 2000). Scientists and environmentalists coalesced around the need for cutting greenhouse gas emissions. However, the often-overlooked policy response was to turn the issue into a scientific one with substantial investments in scientific research. This settlement suited both scientists, who were eager to narrow the uncertainties in their models, and policymakers, for whom it provided an excuse for inaction (Sarewitz and Pielke 2000).

Some, including fossil fuel companies and those on the political right that accepted the reality of anthropogenic climate change, tried to sever the link between the problem of climate change and what they saw as an ideological crusade. Mitigating and adapting to climate change was, for some, an 'engineering problem' with 'engineering solutions'.[3] One can see how the move to geoengineering as just another 'straightforward engineering problem' (Levitt and Dubner 2010) becomes relatively frictionless. The wider point is that, along with the scientising of the problem of climate change, we are seeing a growing scientising of its suggested solutions.

Climate change as experiment

The science of climate change predates its political significance by decades. The idea of planetary experimentation runs through this history, although its meaning shifts over time. Scientists initially saw climate change as a profoundly interesting natural experiment. Later, however, the idea of experimentation is used to highlight the uncertainties, surprises and responsibilities that come as humanity's role is elucidated.

In a 1939 paper on the composition of the atmosphere through the ages, the British amateur scientist Guy Callendar described carbon dioxide emissions as a 'grand experiment' (Harper 2009). In the late 1950s, Roger Revelle, who would go on to influence Al Gore's climate activism, claimed that 'human beings are now carrying out a large scale geophysical experiment of a kind that could not have happened in the past nor be reproduced in the future'. According to Spencer Weart (2008, p. 29), this was not a political point, but a scientific one. Revelle was intensely interested in what was happening to the planet (Weart 2008). He proposed a monitoring station on Mauna Loa in Hawaii to keep track of atmospheric carbon dioxide (Edwards 2001), which has in the years since produced the 'Keeling curve', charting the climb towards 400 parts of carbon dioxide per million in the atmosphere by 2014.

The language of experimentation reinforces both political and scientific arguments. Laboratories and experiments are also invoked to emphasise the need for both scientific understanding and scientific *control* of the grand experiment. Climatologist Stephen Schneider, in making the case for Earth systems science in a book called *Laboratory Earth*, repeats the conventional argument that 'much of what we do to the environment is an experiment with Planet Earth, whether we like it or not' (Schneider 1997, p. xiv). This in his view makes it irresponsible to ignore our new powers of prediction: 'It is no longer acceptable simply to learn by doing. When the laboratory is the Earth, we need to anticipate the outcome of our global-scale experiments before we *perform* them' (Schneider 1997, p. xii, italics in original).

Schneider, like many climate scientists, expresses exasperation at the lack of international action to mitigate climate change. Pointing to what he regards as the relatively minor uncertainties inherent in current understandings of climate, he concludes that regardless of any substantial political action, 'Laboratory Earth continues to grind out the answer – experimentally . . . a transient experiment' (Schneider 1997, p. 88).

For Schneider, the experimental condition of the planet is an argument for action, but not for a technological fix. Schneider was, before his death in 2010, a prominent critic of geoengineering. Similarly, Al Gore (2009) argues against geoengineering by saying that 'we are already involved in a massive unplanned planetary experiment . . . We should not begin yet another.' But for others this sort of analysis easily leads towards geoengineering. Our growing knowledge of the planetary experiment leads some to conclude that geoengineering is acceptable or even humanity's responsibility. James Lovelock (in Fleming 2010, p. 228) has argued that 'we became geoengineers soon after our species started using fire for cooking'. For Stewart Brand (2010), geoengineering is a mere extension of gardening. Brand's slogan at the start of his *Whole Earth Catalog* in 1968 – 'We are as gods and we might as well get used to it' – soon morphed into 'We are as gods and we might as well get good at it' for later editions of the book. T. Nordhaus and Shellenberger (2007, p. 135), in their critique of an environmental 'politics of limits', produce a challenge for 'post-environmental' politics: 'The issue is not whether humans should control Nature, for that is inevitable, but rather how humans should control natures – nonhuman and human.' In 2013, Nordhaus and Shellenberger's *Breakthrough Journal* published a piece arguing that it is time to 'embrace geoengineering', as though it were an off-the-shelf technology (McInnes 2013).

According to Bruno Latour (1999, p. 43), 'for the world to become knowable, it must become a laboratory'. More recently, Latour (2013) has concluded that the changing climate is a 'laboratory in which the experimenters are imprisoned'. One might imagine that the problem of climate change demands a variety of experimentation and innovation in politics, civil society and business, as well as science and technology. But climate change has instead become depoliticised (Badiou 2008, p. 139). We have been paralysed rather than scared into action (Beck 2010). The adage is that 'climate is what you expect; weather is what you get', but as our everyday meteorological experience gets subsumed under an overarching global science, 'gradually, weather has become climate' (Urry 2011, p. 18). The search for a global and direct solution, informed by universal scientific understandings, has overshadowed local actions that may tackle climate change in orthogonal ways. The construction of climate change as a scientific and political experiment has in turn put enormous pressure on scientists and their tools.

Climate models have come to attain huge importance in climate science and climate policy as a simulacrum of the present and future world. But their apparent political weight belies the weightlessness that makes them scientifically attractive. They are, as Johann Feichter argues, a virtual reality as well as a 'virtual laboratory' (Feichter 2011, p. 216). The data and observations of climate science are necessarily incomplete, but the models allow scientists to construct workable worlds with which to experiment.

In the eighteenth and nineteenth centuries, meteorology was a 'descriptive science' (Gramelsberger and Feichter 2011) or, as physicist Theodore von Karman called it, a 'guessing science' (quoted in Harper 2006). The twentieth century saw the introduction of thermodynamics and hydrodynamics into meteorology, along with the computing power to construct the first general circulation models of the global climate. But these theoretical developments needed testing. In the absence of a real world accessible on the required geographical and temporal scales, computer models became the site for meteorological 'experiments'. In this way, meteorology could start living up to the dream of Vilhelm Bjerknes that it would become a 'physics of the atmosphere' (Gramelsberger and Feichter 2011), able to predict the weather as accurately as the movements of the planets.

Since the 1970s, layers of resolution and complexity have been added to climate models. First, models of the land were appended to those of the atmosphere. The oceans and sea ice were added in the 1990s. More recently, aerosols and the carbon cycle have been included (IPCC 2001). The current enthusiasm is for 'Earth system models' that include air chemistry and plants and 'integrated assessment models' that introduce humans, with consideration of economics, technology and other factors. These have allowed economists to calculate, with no small controversy, economically 'optimal' levels of atmospheric carbon dioxide (e.g. W. D. Nordhaus 2007).

As climate science has become intertwined with climate politics, climate model experiments have been called on not just to represent the climate, but to project likely climates under different scenarios. So the level of atmospheric carbon dioxide in the model can be doubled or trebled as the experiment runs and the future climate is projected. These experiments are sometimes referred to as forced, because of the additional 'radiative forcing' applied to the model. The Intergovernmental Panel on Climate Change (IPCC) now distinguish between 'prediction', which is seen as neutral, certain and therefore a hostage to fortune, and 'projection', which is more obviously a product of models with their myriad uncertainties and the potential for altered inputs (Bray and von Storch 2009). 'Prediction' tries to account for the uncertainties in both from the initial conditions and the model itself, whereas 'projection' adds the long-term uncertainties that may come from human agency and choices made about mitigation.

Climate model experiments have taken place for as long as climate models have existed (Fleming 2010). The first successful attempt to model climate in a computer was known as 'the first experiment' by its orchestrator, Norman Phillips (Lenhard 2007). Experiments are a way for models to 'talk back' to the world (Hastrup 2013, p. 5), a vital bridge between theory and application (Varenne 2001; Gramelsberger and Feichter 2011). But as I discuss in Chapter 6 the task of authoritatively projecting future climates is complicated by these experiments with models being at the same time experiments *on* the models.

Global models for climate change have become a 'vast machine' (Edwards 2010), employing thousands of researchers and huge computing resources. The overarching fact of global climate change has remained stable for decades, but at times it feels politically precarious. Scientists are often overly defensive in the face of organised scepticism (in the political, rather than the Mertonian, sense) and policy indifference. A large part of the rationale behind the creation of the IPCC was to build public and political trust in climate science, and the institution has become extremely protective of its cognitive authority (Grundmann 2007). This has created a situation in which climate scientists are compelled towards ever-greater detail, precision and certainty in order to maintain their consensus and in the false hope that this will be politically decisive (see Sarewitz 2004).

(The classical allegory might be the trial of the Danaids, who were destined to spend eternity attempting to fill a leaking vessel.) The models can't be strictly verified or validated against the real world because Nature is not a closed system and the models don't give single answers (Oreskes *et al.* 1994). The models may 'ring true' (Oreskes *et al.*), but they are not a 'truth machine' (Wynne 2010, p. 296).

The connection between climate science and policy is not linear. The blending of science and policy in institutions like the IPCC is itself an experiment (Wynne 2010). However, the particular way in which climate science and climate politics have developed their relationship helps to explain why geoengineering is seen to mark such a dramatic break. The assumption is that a fragile political consensus rests on a scientific consensus that is the product of modelled results rather than palpable experiments and observations. This has prompted many scientists and environmentalists to push wholeheartedly for a global agreement on cutting emissions, using market mechanisms to increase the cost of emitting. Any alternatives are seen as a step backwards. Geoengineering, as currently imagined, promises to sever the connection between problem and solution upon which so much effort has been expended. For those with long histories in the climate debate, geoengineering threatens a sort of gestalt shift. However, the science and its uses have a more complicated relationship.

Instrumental climates

The delineation between understanding, prediction and control in climate and weather science is further complicated by the growing range of uses to which expertise and data are put. In the first half of the twentieth century, insurance companies were already using weather forecasts in insurance policies for sporting events (Harper 2008, p. 6). In a more recent twist, meteorology has become a tool for financial innovation through 'weather derivatives', deepening a longstanding relationship with private industry (Randalls 2010). With the professionalisation of meteorology, the science went beyond being merely observational to take on a role as a risk manager. The growing confidence of the discipline meant that in a sense, scientists in effect began to assume responsibility for the weather through institutions that would become increasingly scientific. In the UK, the Met Office took on this role, particularly with regard to extreme weather events (Hall 2012).

Social scientists have pointed out that our relationship with extreme weather is far from straightforward. According to Hilgartner (2007, p. 153), in advanced, rich countries, 'there are no natural disasters, only sociotechnical ones'. The point is that the experience of an extreme weather event such as Hurricane Sandy or Hurricane Katrina depends more on a society's resilience or vulnerability than on meteorological conditions. Even if we accept that the risks, whether from weather or seismic activity, cannot be perfectly predicted, we can follow Mary Douglas (1992) in concluding that risks are not objective; they are assessed and chosen according to cultural norms. So societies can and should accept some responsibility for natural disasters. They are never just 'acts of God' (Steinberg 2006).

With the growth of climate science, the calls for improved weather event attribution have become louder. The reasoning is that more sophisticated models should be able to calculate the probability that weather events are attributable to climate change rather than being just the vagaries of normal weather. As well as connecting the long-term, abstract climate debate with people's lived experience, event attribution would also make targeting resources for climate adaptation more precise. Hulme and colleagues see dangers in the creep of this 'predict-and-adapt paradigm' (Hulme *et al.* 2011, p. 764). As well as ignoring the point made above about the unnaturalness of 'natural' disasters, this approach assumes that current uncertainties are tractable. Even if uncertainties can be resolved to the satisfaction of climate modellers, they will be contested and prised much farther apart in the public domain (Hulme *et al.* 2011; also Stilgoe 2007). Hulme and colleagues conclude that this overreaching of climate modellers 'politicizes climate science [and] scientizes adaptation politics' (Hulme *et al.* 2011, p. 765).

All of this means that the discussion about liability that a geoengineered world would inevitably demand is already beginning. Every large storm resurfaces a debate about whether it can be blamed on climate change, and if so, on the largest polluters, whether these are felt to be industrialised economies or fossil fuel companies. We might expect such discussions to intensify if we move towards intentional climate change, but there are lessons here, too. Fleming (2010) describes how rainmakers and weather modification researchers attracted the attention of lawyers who could anticipate the challenges of attribution and liability. He goes on to mention rumours that circulated, following the 1952 flood in Lynmouth, North Devon, about cloud-seeding experiments conducted by the Met Office and Ministry of Defence. A small number of local residents believed that the government was to blame for the rainfall that swept away much of their seaside town.

Responsibility in the Anthropocene

> *'My dream holiday is to get in a boat and start from one end, go to the other end of a river and never see anybody apart from the people I've chosen to go with . . . I get to these places . . . and I think "this is no longer natural because somebody has geo-engineered the place". You know, there's a hundred million years of geology and in the blink of an eye we've fucked it up and now the sky's a slightly different colour and actually nowhere is wild; nowhere is natural. And then I realise that that's a complete fallacy . . . Just because you can't see carbon dioxide it doesn't mean we haven't already had a profound change. And so our relationship with nature is rather complicated and very, very difficult to conceptualise.'*
>
> (A scientist working on SPICE)

Scientists have sought new ways to represent the rapid disjuncture in the life of planet Earth caused by the arrival and industrialisation of humans. For some scientists and environmental campaigners alike, climate change makes complete the end of Nature. Bill McKibben (1989) has argued that 'we have changed the atmosphere, and thus we are changing the weather. By changing the weather, we make every spot on earth man-made and artificial.' More recently, some scientists have begun to talk about 'the Anthropocene' as a new phase in the Earth's geology. At the time of writing, the Anthropocene Working Group of the International Commission on Stratigraphy, who pronounce on such matters, has yet to report. But this line of thinking has already captured attention. The front cover of *The Economist* in May 2011 announced 'Welcome to the Anthropocene.' Scientific papers have determined the arrival of a new epoch (e.g. Steffen *et al.* 2012). Where we previously thought we lived in the Holocene interglacial, a period with a calm, habitable climate, we are now told that a Rubicon has been crossed. The Anthropocene is imagined to bring new instabilities. Geology, once thought to pass humanity by without breaking its slow step, would now appear to be under our influence.

The term 'Anthropocene' was introduced by Paul Crutzen at the turn of the millennium, writing with Eugene Stoermer (Crutzen and Stoermer 2000). Crutzen, who would turn his attention to geoengineering a few years later, is a Dutch atmospheric chemist. In 1995 he won a Nobel Prize for his research on ozone, revealing dangers from chlorofluorocarbons that were rapidly tackled using an international agreement – the Montreal Protocol of 1987 – that scientists still hold up to highlight the possibilities of science-based policy. Crutzen possesses the extended public authority that comes with any Nobel Prize, but especially with one in a policy-relevant area of science.

Since the 2000 paper, there has been a proliferation of visualisations of the Anthropocene. Photos of the Earth at night with its visible cities contrast with the 'pale blue dot', Carl Sagan's phrase capturing the fragility and loneliness of the planet (Jasanoff and Martello 2004). Scientific papers show a range of graphs marking the disjuncture. Human population, gross domestic product, fertiliser consumption, water use and numbers of motor vehicles and McDonald's restaurants (Steffen *et al.* 2012) all show the hockey stick shape, ticking upwards in the second half of the twentieth century, and are taken as indicators of human influence.

There is a vociferous scientific discussion about the start date of the new era, with some arguing for the Industrial Revolution (Crutzen and Steffen 2003) and others for the Neolithic spread of agriculture thousands of years earlier (Ruddiman 2003; see also Hamilton 2013). Baskin (2014) makes the point that because the term is a hybrid scientific–political one, the community around the idea *need* the Anthropocene to be current and sudden. Regardless of the precise start date of the Anthropocene, these ideas have important connections with politics. The Anthropocene, though it purports to be a geological concept, is as much a vision of our future (Scott 2013). The Anthropocene is a scientifically constructed mirror with which humanity is expected to rethink its responsibilities.

Humans would seem to have broken down the wall between natural history and their own history (Chakrabarty 2009, p. 201). The Anthropocene means admitting a new degree of ownership and stewardship. The Anthropocene casts these responsibilities in scientific terms, even though human beings are hard to fit into scientific frames. It is scientific knowledge of our impact on the planet that separates negligence from recklessness. The argument runs that we can no longer excuse our actions. Andy Revkin (2011) puts it like this: 'It was easier to be in a teen-style resource binge before science began to delineate an edge to our petri dish. We no longer have the luxury of ignorance.' The language of the Anthropocene has drawn close together previously disparate questions of understanding, responsibility and control as applied to the Earth. Andrew Mathews (2011) argues that the recent reawakening of geoengineering ideas has been encouraged by the 'cosmopolitics' of the Anthropocene and the scientific imagination of climate change at a global scale. Geoengineering represents a particular way of scientists' taking responsibility for what we know about climate change, in the light of policy recalcitrance on mitigation. The anthropocentrism of the Anthropocene provides would-be geoengineers with a rationale, and it forces even those who blanche at the idea of geoengineering to consider humanity's responsibility for a changing climate. Perhaps we should not be surprised therefore to find Anthropocene writers arguing that 'many approaches could be adopted, ranging from geo-engineering solutions that purposefully manipulate parts of the Earth Syst… ing active stewards of our own life support

…ocene' term – a set of technical and politi… …uage – has allowed it to gain traction and …arth system scientists. In doing so, it eased … that of planetary boundaries.

…olicy offshoot of Earth system science. A …tröm, but whose authorship substantially …ene ideas, announced the project to iden… …ting space for humanity' (Rockström *et al.* …tem processes to provide the dimensions …d phosphorous cycles, ozone depletion, … alongside climate change. Their assess… … nitrogen use, climate change and biodi… …led.

… attracted a good deal of scientific criti… … Schlesinger (2009) argued that limits and clear lines (between success and failure; life and death) may provide some comfort, but the time spent arguing about them could be better spent taking action even if we are uncertain. Myles Allen (2009) saw the multiple boundaries as a distraction from efforts to hold global warming at two degrees Celsius. They and the other respondents went on to deconstruct the scientific basis for the suggested boundaries, arguing that they are too high, too low, impossible to calculate or incomplete. In a separate intervention, Ted Nordhaus and Michael Shellenberger, joined by Linus Blomqvist, deconstructed the evidence behind the proposed boundaries and their suggested limits (T. Nordhaus *et al.* 2012). (Nordhaus and Shellenberger are self-professed 'ecomodernists' who regard the framing of environmental issues in terms of 'limits' as part of the problem.)

The original paper restrained itself to scientific claims. Subsequent publications from its authors and their collaborators have been more politically expansive. According to one piece, the aim with planetary boundaries thinking is to identify 'nonnegotiable limits' (Steffen *et al.* 2011b). Speculating on the political conditions for governing the boundaries, the authors conclude the following:

> Ultimately, there will need to be an institution (or institutions) operating, with authority, above the level of individual countries to ensure that the planetary boundaries are respected. In effect, such an institution, acting on behalf of humanity as a whole, would be the ultimate arbiter of the myriad trade-offs that need to be managed as nations and groups of people jockey for economic and social advantage.
>
> (Steffen *et al.* 2011b).

For Roger Pielke, Jr (2013), this goes beyond scientism; it is an anti-democratic 'power grab'. Others see the potential for closing democratic debate as more subtle. Latour (2013) sees a switch taking place. Where we once saw the planet as immobile and politics as potentially revolutionary, we now see politics as hopelessly static in the face of a planet that is able to take us by surprise in the speed of its glacier retreats, sea-level rises and extreme weather. The brinkmanship of the planetary boundaries approach if these boundaries are seen as scientific facts rather than political limits is likely to lead to fatalism rather than action.

The planetary boundaries would seem to be the latest incarnation of a growing scientific exasperation with 'an inconvenient democracy' (Stehr 2013). James Lovelock has been most explicit, arguing that our survival in the face of climate change demands 'an unusual degree of human understanding and leadership and may require, as in war, the suspension of democratic government' (Lovelock 2009, p. 61). But the increasing volume of discussions about climate tipping points, climate emergencies and the framing of geoengineering as a response (Markusson *et al.* 2013) reflects similar sentiments.

The imagination of planetary boundaries in the Anthropocene forces the issue of necessary action on climate change, but such ideas rarely specify what should be done. To understand how the peculiar idea of stratospheric particle injection, which would lead the debate about geoengineering, found favour, we need to turn to an unlikely source of inspiration.

Volcanic experiments

As I explained in Chapter 1, the short history of geoengineering research contains very few tangible experiments. But an important natural experiment (Morgan 2013) took place in 1991 with the eruption of Mount Pinatubo in the Philippines. Pinatubo threw millions of tonnes of ash, ice and sulphate up to the stratosphere in less than a day (Guo *et al.* 2004). The scale of Pinatubo's eruption is dwarfed by earlier events such as at Mount Tambora (Briffa *et al.* 1998), but it was large enough to have a demonstrable impact on the global climate and took place at a time in history when it could be measured, monitored and modelled by the machinery of late-twentieth-century global climate science. As well as generating death, devastation and disruption, Pinatubo contributed evidence in support of a scientific idea that had been circulating for more than a decade.

The idea of stratospheric particle injection began with Mikhail Budyko in 1974. Others had observed previously the cooling effect of large volcanic eruptions, but Budyko provided early calculations of the relatively small quantities of sulphur required to engineer this effect. He also provided the first example of the trope that injecting sulphur dioxide into the stratosphere was cheap and easy.

> Since modern high-altitude aircraft carrying a load of about 15 tons can reach the level of the aerosol layer, this mass of reagent can be transported to the lower stratosphere by several aircraft operating every day equipped with a device for burning sulphur in the atmosphere . . . This mass is 10^{-4} [one ten thousandth] of that due to man's activity, which, according to contemporary data, constitutes hundreds of millions of tons per year . . . Obviously, such amounts are not at all important in environmental pollution.
>
> (Budyko 1977, pp. 240–241)[4]

As an alternative to aeroplanes, Budyko also suggested guns or missiles to fire the sulphate to the required level. The particles would need to be high enough to enter the stratosphere, at which point they would spread around the globe rather than falling rapidly back down to Earth. In the troposphere, the lower part of the atmosphere, particles return to the surface of the Earth in a matter of days. The stratosphere is 'convectively stable', which means that sulphate particles can stay aloft for years. If a large volcano erupts in the tropics, a mechanism known as Brewer–Dobson circulation can take the detritus to the poles and spread it around the planet within days.

The idea of stratospheric particle injection and estimates of its intended and unintended consequences are bound up in the science of volcanoes. The eruption of Pinatubo had a discernible and dramatic effect not just on global temperatures, but also on patterns of rainfall around the world (Trenberth and Dai 2007). Volcanoes provide not just a natural analogue for the effects of geoengineering, but also an indication that the planet could be cooled with relatively small quantities of particles (only a few million tonnes for the whole planet): the ultimate high-leverage technology.

Pinatubo has become a muse for stratospheric particle injection research. As has been discussed elsewhere (Hamilton 2013; Stilgoe *et al.* 2013) the climatic disruption caused by earlier volcanic eruptions has inspired cultural as well as scientific insight. The red sky behind Munch's open-mouthed man in *The Scream*, a painting that reflects humanity's fraught relationship with Nature, was purportedly an after-effect of the eruption of Krakatoa in 1883.[5] The largest eruption of modern times, Tambora in 1815, led to the 'year without a summer', inspiring Mary Shelley (then Mary Godwin) to write the definitive parable of humanity's relationship with Nature and technology, *Frankenstein; or: The Modern Prometheus*. According to a recent book on the cultural impact of this eruption by Gillen D'Arcy Wood (2014), painters such as Turner and Constable used unusual quantities of red paint during this period as they tried to reproduce the dusty sunsets of the time. The enthusiastic economists behind *Freakonomics* regard Pinatubo as a cornucopia of 'positive externalities'. They point not just to the global cooling but also to improved plant growth from diffused sunlight and 'some of the prettiest sunsets that people had ever seen' (Levitt and Dubner 2010, p. 176).

Research on volcanoes has given credence to an idea that might otherwise have seemed hopeless: creating a sunshade for the Earth. And while some popular representations of geoengineering, such as that provided by Levitt and Dubner (2010), suggest that the task is merely to engineer the same thing, the construction of a credible research agenda and community for geoengineering demands more than just a volcano.

Rehabilitating geoengineering

David Keith had noted in 2000 that geoengineering had fallen out of fashion (Keith 2000). He had himself taken what he described as 'a serious look at geoengineering' in 1992 (Keith and Dowlatabadi 1992), around the time of the National Academy of Sciences study on greenhouse warming (NAS 1992) and a 1996 special issue of *Climatic Change* that followed a symposium at the American Association for the Advancement of Science (see Marland 1996). This special issue established a long-standing pattern of multidisciplinarity, putting climate science alongside law, ethics and economics in assembling relevant expertise for understanding geoengineering. The papers in this special issue all acknowledge, implicitly or explicitly, the taboo shrouding geoengineering. Economist Thomas Schelling noted that geoengineering had become 'unmentionable' until the NAS had begun to weigh up the options (Schelling 1996, p. 303). Most natural scientists at the turn of the millennium could relatively easily express distaste at geoengineering or dismiss it entirely as a half-baked relic of Cold War techno-enthusiasm. Most of the scientists who did write about geoengineering did so with extreme caution in their framing and the nuances they presented.

According to the conventional narrative, the taboo on geoengineering was broken in 2006 with the publication of a paper by Paul Crutzen. Crutzen was less cautious. He expressed enthusiasm for the idea of SRM and drew an explicit link with the lack of political action on carbon emissions. For Crutzen, the problem of climate change was an undisentangleable Gordian knot, and geoengineering provided a sword or, as he put it, 'a contribution to resolve a policy dilemma' (Crutzen 2006). This was not Crutzen's first statement on geoengineering – he had suggested it as a possible offshoot of his own Anthropocene thinking (Crutzen 2002) – but the 2006 paper became the landmark. Crucially, Crutzen could not be dismissed, as Edward Teller and Lowell Wood could be, as Dr Strangeloves with hammers looking for nails. Crutzen was a Nobel Prize-winning environmental scientist with an impeccable pedigree.

Scientists around Paul Crutzen at the time recognised the power his comments would have, with some trying to persuade him not to publish. Stephen Schneider, the founding editor of the journal *Climatic Change*, agreed to publish the paper, but only if it was wrapped in commentary from various other authors. Mark Lawrence, who had been involved in similar discussions about the promises and threats of ocean iron-fertilisation (OIF) experiments, discussed the politics of the scientific community's wholehearted engagement in geoengineering (Lawrence 2006). Ralph Cicerone pointed to a lack of knowledge, concluding in his commentary that 'refereed publications that deal with such ideas are not numerous nor are they cited widely' (Cicerone 2006, p. 221). The aim of including these pieces was to take some of the magic out of Crutzen's magic bullet, but his paper moved into the climate science community more easily than the counterarguments and modifiers that originally accompanied it.

Scientists with climate models who had previously been uninvolved in geoengineering discussions began to ask themselves whether they should use their tools to work out how a geoengineered climate might look in the future, and a number of analysts from other disciplines joined the party. Cicerone's (2006) conclusion would not hold for long. The number of geoengineering papers being published each year increased from fewer than 20 in 2005 to more than 100 by 2010 (Oldham *et al.* 2014). Stephen Schneider subsequently wrote about the importance of the Crutzen (2006) paper, saying that 'in this case, the messenger is the message' (quoted in Morton 2007). Schneider attributed Crutzen's arguments to 'exasperation with the capacity of society to mitigate the "right way"' (Schneider 2008, p. 3847). Crutzen's own justification suggests a peculiar brinkmanship: 'It was meant to startle the policymakers . . . If they don't take action much more strongly than they have in the past, then in the end we have to do experiments like this' (quoted in Borenstein 2007).

My own conversations with scientists reinforce the importance of the Crutzen paper. It generated a range of responses. One climate scientist told me this:

> *'Paul Crutzen had published his paper saying "hey this is a possibility" and that's really what kicked it all off . . . there were two reactions to that: "gosh if he considers it a possibility then we have to take it seriously" and the other reaction was "my goodness, what does he think he's doing?"'*

SRM promised to sever the link between greenhouse gas emissions and climate change, upon which so much scientific research and political negotiation had depended for so long. For a prominent environmental scientist to be openly discussing it as an option was radical. Another scientist told me that his own interest was

> *'triggered . . . by the Crutzen paper in 2006 and I remember a discussion at our institute . . . very close to the time of the paper coming out . . . Some people said*

> *"Well it's our area of expertise" . . . But then there was also a strong group of people saying that we cannot do this . . . just from a gut feeling I guess: you should not tinker with nature and the risks are too high . . . then there was this slippery slope argument: we should not raise the topic at all because just giving the impression that there may be a technical fix to the climate problem could lower the interest in doing some real mitigation . . . But, well, at some point I just thought "well, it's too late".'*

This researcher's final point captures a widely held view that the Crutzen (2006) paper had let the cat out of the bag. To many scientists, it seemed as though the taboo that surrounded geoengineering had been lifted.

A taboo, however, implies protection from an unspeakable truth. It suggests there is a technology ready to be unleashed on the world. This had been the way in which stratospheric particle injection, Crutzen's proposal of choice, had historically been discussed, dating back to Budyko's (1974) original calculations. Budyko had suggested that geoengineering was straightforward: 'we need only deliver a fixed amount of sulphur dioxide to the level of the aerosol layer . . . transported to the lower stratosphere by several aircraft operating every day equipped with a device for burning sulphur in the atmosphere' (Budyko 1974, p. 240). Crutzen concludes that 'provided the technology to carry out the stratospheric injection experiment is in place, as an escape route against strongly increasing temperatures, the albedo adjustment scheme can become effective at rather short notice' (Crutzen 2006, p. 216). The language here is interesting. There is confidence that the scheme 'can become effective', and yet this is also an 'experiment'.

Following Paul Crutzen, David Keith and others, geoengineering became, in the words of Alex Steffen (2008), 'a set of 20th century proposals kitted out in 21st century drag'. The tools of research had become hugely sophisticated, but the imagined geoengineering technologies remained resolutely low tech. The task of civilising geoengineering, detaching it from the huckster rainmakers and hawkish technological fixers who were its previous owners, meant reconnecting it with climate science. In doing so, the technologies themselves were left largely overlooked.

Self-governance

The Crutzen (2006) paper created large ripples in the small community of researchers looking at geoengineering. Many of them had previously been involved in the climate change debate and had been part of the collective voice of science calling for urgent action to cut greenhouse gas emissions, through the IPCC and other channels. Some, such as Alan Robock, had also been involved in debates about nuclear disarmament. Robock and another early geoengineering researcher, Mike McCracken, had conducted influential studies on the climatic effects of nuclear war, modelling the so-called nuclear winter that would cool the Earth as fire and explosions launch smoke and debris into the air. This connection goes some way to explaining why Robock chose to publish an influential scientific critique of geoengineering in the *Bulletin of Atomic Scientists*, a journal that had over decades warned of the dangers of nuclear weapons and whose cover famously features a clock, the hands of which point closer to midnight as the threat of annihilation builds.

Robock's (2008) article refers to the Crutzen (2006) paper, which he described as 'controversial', and to a piece by Tom Wigley (2006) published in *Science* the same year. Robock offers a 'fairly comprehensive list of reasons why geoengineering might be a bad idea' (Robock 2008, p. 14). The tone is not one of offering new research findings but one of scientific self-policing. He is reminding his scientific colleagues of what is already known about the hazards, lest their enthusiasm should get the better of them. Although he summarises the science that points to possible unintended consequences, his list includes ethical and political concerns, too, such as the irreversibility that comes with technological dependence, military and corporate involvement, and the challenge of getting agreement on an ideal planetary temperature. Robock ends his list by adding that in addition, 'there is reason to worry about what we don't know' (Robock 2008, p. 17).

Robock's (2008) paper is a prominent example of scientific leadership of a debate about ethics and governance. It continued the multidisciplinary approach to geoengineering assessment that had quickly become familiar, while still being notable. (His piece sat alongside a shorter commentary from his colleague Martin Bunzl, a philosopher.) Two years later, a group containing most of the world's active geoengineering researchers was brought together to try to discuss norms and rules for what was already becoming a sub-field of research. For their 'Asilomar moment' (Schäfer and Low 2014), geoengineering researchers visited the location of the 1975 conference at which the nascent genetic engineering community had gathered to discuss the dangers posed by their newfound ability to engineer bacteria. One of the instigators of the original conference at Asilomar, Paul Berg, joined the organising committee of the geoengineering version.

The 2010 Asilomar conference was chaired by Mike McCracken and included a wide range of expertise and interest. Nevertheless, the conference had the same tension as its earlier counterpart, caused by multiple ideas of responsibility in geoengineering research. The 1975 conference and the short-lived moratorium that the community had adopted while waiting for it to conclude are often held up (including in the publicity material for the geoengineering meeting) as an example of the scientific community identifying problems early and taking responsibility. But accounts from the time (Rogers 1975) and subsequently (Nelkin 2001; S. Wright 2001) suggest that many of the participants were more motivated by a desire to head off top-down regulation with a promise of self-governance. Geneticist Stanley Cohen said at the time, 'If the collected wisdom of this group doesn't result in recommendations, the recommendations may come from other groups less well qualified' (quoted in Nelkin 2001). Although the discussion was motivated by non-scientific issues, its framing was narrowed to technical considerations in ways that, as Sheldon Krimsky (2005) argues, have continued to define regulation of genetic technologies. The discussions revealed vast uncertainties, but the dominant assumption was that research should be allowed to proceed in order to clear these up and realise the vast imagined benefits of technology. Susan Wright (2001) also described the original Asilomar conference as reinforcing the myth that such scientists can ever be 'self-governing'. Their work is governed by external pressures and expectations, such as, in the case of genetic engineering, a hungry biotechnology industry, as well as internal norms, rules and ideals.

The 2010 Asilomar moment was more multidisciplinary. There was also evidence that some scientists' concerns extended beyond merely the legal or technical, to encompass what might be called 'second-order reflexivity' (Schuurbiers 2011) or the 'governance of governance'. David Keith and Ken Caldeira both publicly took issue with the organisation of the meeting and its apparent connections with a company that had sought to make money from an OIF scheme. Keith chose to attend the meeting to make his views known, while Caldeira sat it out.

In the young world of geoengineering, Ken Caldeira's decade-long string of scientific publications makes him an old hand. He is an outspoken climate scientist with a focus on ocean acidification, and he is fond of expressing climate policy arguments in terms of stark ethical choices. He is a former colleague of Lowell Wood's at Lawrence Livermore National Laboratory who began modelling solar geoengineering at the end of the twentieth century because he was convinced it would wreak havoc with the planet. The way he tells the story, he was surprised that the climate models seemed to disagree. He subsequently became an important midwife for the nascent geoengineering research community, although he identifies himself more broadly as an atmospheric scientist.

Caldeira's and Keith's concerns, and the contributions of the varied participants, challenged the assumption held by some other participants that the science should be autonomous from outside interference, a view that would continue into discussions of the governance of outdoor experiments that I discuss in Chapter 5. The report of the meeting (Asilomar Scientific Organizing Committee 2010) was predictably light on tough, agreed rules for responsible research. The meeting was criticised by those who didn't attend, who were understandably suspicious of an attempt to circumvent stringent rules (e.g. ETC Group 2010). And it was lauded by some who were there, such as science writer Jeff Goodell, who 'witnessed the birth of something new – call it the conscience of a geoengineer' (Goodell 2010). For most people there, including at least five participants from the Royal Society's study that I describe in the next chapter, the meeting represented neither of these things. But it was nevertheless a prominent experiment in responsibility. The agenda and participants of the meeting were markedly broader than the heavily scientific 1975 Asilomar conference.

CDR vs SRM

In its rehabilitation, geoengineering has held together some ideas and technologies that on the face of it would seem to have little in common. The distinctions made within and around the label 'geoengineering' suggest certain interests, although they are by no means fixed. The most common separation is between ideas of SRM, which are the primary focus of my analysis, and those of carbon dioxide removal (CDR).

It is worth noting here that the label 'solar radiation management', described by one leading climate scientist as 'positively Orwellian',[6] was coined as a joke by Ken Caldeira, who was attempting to lampoon bureaucratic language.[7] This joke, however, stuck, and one could argue that the simplicity of the label and others such as 'energy balance' exemplify the reductionism of geoengineering research. 'Climate control', which one might admit was the more honest way to talk about the target of geoengineering, seems substantially harder to do and to model than the task of rebalancing the energy hitting and leaving the planet's surface.

It is SRM that has attracted most attention because of the potential for rapid climatic change that is imagined. Interest in space mirrors (e.g. Angel 2006) has largely faded as interest has grown in stratospheric particle injection and marine cloud brightening (e.g. Latham 2002), which are seen as more feasible. CDR aims to treat the cause rather than the symptoms of climate change. Advocates of approaches that fall into this category, including machines for direct air capture (artificial trees), biochar (the burning and burying of plants), conventional carbon capture and storage and planned tree growing, often express their distaste at being included alongside what they see as the high-stakes, unproven and potentially catastrophic ideas of SRM. These CDR approaches, on the whole, are proven to do their job at small scale but are beset by practicalities and vast costs when put up against the planet's carbon dioxide levels. They are, in the main, direct, brute-force approaches to the problem of climate change. The SRM proposals that have come to dominate geoengineering discussions are as yet unencumbered by practical considerations, remaining in the world of hopes and fears.

Drawing a distinction between SRM and CDR and making certain suggestions for research and governance suit some researchers, as do distinctions between 'hard' and 'soft' geoengineering (Olson 2012), but these lines are not clear. There are proposals for incremental, bottom-up SRM, such as whitening the roofs of buildings, which seem mundane. And the claims offered for OIF, which I will discuss more in Chapter 5, suggest that even though the aim is to remove carbon dioxide, the idea is to use a small ecological lever to tip an ecosystem into a dramatically different state.

My aim here is not to attempt to clarify definitions or draw new lines. Indeed, attempts to do so should be carefully scrutinised according to whose interests they serve. Others have offered some distinctions that may enable improved discussions of governance. Steve Rayner has offered a distinction between 'black-box engineering' and 'ecosystem enhancement' which would lump OIF and stratospheric particle injection together as ideas of concerns (Rayner *et al.* 2013). Ben Hale (2013) has suggested that we should look at geoengineering ideas according to their intent, separating those that look to remediate from those that intentionally take the climate to a new state. The idea of remediation is itself problematic, as I discuss below. But the notion of intent is hugely important to geoengineering discussions, as will become clear with the discussion of SPICE (Stratospheric Particle Injection for Climate Engineering) in later chapters. Although some of the scientists involved in geoengineering are occasionally uncomfortable defining themselves by their intentions, David Keith has spoken forcefully against the conflation by Lovelock and others that makes the experiment of climate change equivalent to the experiment of geoengineering, as discussed earlier. For Keith, the Earth at the moment is not 'engineered', because 'pollution is not engineering. Intent matters' (Keith 2010, p. 27). Keith and others delineate geoengineering proper from the mass of pollutants that are already inducing various warmings and coolings in the climate system.

The pseudoeconomics of geoengineering

Central to the success of its rehabilitation is what Scott Barrett (2008) calls 'the incredible economics of geoengineering'. Looking at the political economy of climate policy, Barrett concludes that the incentives for mitigation look so feeble compared to those for geoengineering that its deployment is almost inevitable. Central to his argument, and to the geoengineering debate more broadly, is the prevalent but flawed assumption that, while CDR technologies, some of which are well advanced and well understood, look breathtakingly, perhaps prohibitively, expensive, SRM would be cheap. Like a planetary vaccine, a tiny injection would have a profound effect. David Keith says, 'This near million-to-one leverage is at the root of both the risk and the promise of stratospheric aerosol geoengineering; it is the underlying reason why it is such a powerful and frightening tool' (Keith 2013, p. 67).

The 1992 National Academies study had come to a similar conclusion: 'one of the surprises of this analysis is the relatively low costs at which some of the geoengineering options might be implemented' (NAS 1992). Keith (2000, p. 263) had previously said that 'it is unlikely that cost would play any significant role in a decision to deploy stratospheric scatterers because the cost of any such system is trivial compared to the cost of other mitigation options'. These calculations were repeated in cartoon form in the popular book *Superfreakonomics* (Levitt and Dubner 2010). As with its predecessor, *Freakonomics*, this book's exaggerated posture is to strip the morality away from decision-making to get at an underlying economic truth: 'Morality, it could be argued, represents the way that people would like the world to work – whereas economics represents how it actually does work' (Levitt and Dubner 2005, p. 13). Their calculations of the realpolitik of geoengineering reveal to them that 'for anyone who loves cheap and simple solutions, things don't get much better' than stratospheric particle injection (Levitt and Dubner 2010, p. 277).

As I discussed in Chapter 1, early speculation on nuclear power suggested that it would be 'too cheap to meter'. There are few sensible commentators who would share the *Superfreakonomics* view that the affordability of geoengineering makes it the preferred option. Instead, the argument is that it is too cheap to ignore, making it irresistible to desperate countries threatened by climate change or rogue eco-terrorists and therefore too cheap to completely prohibit. Even though they respectively make 'a case for climate engineering' and 'a case against climate engineering', David Keith (2013) and Alan Robock (2014) agree that doing stratospheric particle injection would cost a few billions of dollars per year. For the global economy, such a figure would be 'a mere pittance', according to Victor and colleagues in their dissection of 'science fact' from 'science fiction' (Victor *et al.* 2013).

At no point do these analyses admit the profundity of the technical, social and financial uncertainties. These options are costed as though they exist now. Barrett (2008, p. 45) describes how SRM 'is inexpensive and can be undertaken by a single country, unilaterally', as though a technology were ready to go. For geoengineering, 'cheap' is shorthand used to describe an imagined Promethean, high-leverage technology, and, following dynamics already identified by the sociology of expectations (e.g. Borup *et al.* 2006), neither critics nor cheerleaders have much incentive to offer a more cautious but more realistic price for the technology. Gordon MacKerron (2014) concludes that the vast majority of financial estimates consider only the direct costs of technology, and often do so in highly optimistic terms.

Where serious cost estimates have been performed, the focus has been on the machinery required to, for example, take particles up to a particular altitude (Mclellan *et al.* 2011, 2012; Davidson *et al.* 2012). One such estimate concludes that doing planetary geoengineering would require a fleet of plans 'comparable to the yearly operations of a small airline' (McLellan *et al.* 2011, p. 5). There is little consideration given to the costs of a complete sociotechnical system, although a subsequent paper adds caveats about the uncertainties involved in making such calculations and the 'issues of risk, effectiveness or governance that will add to the costs of solar geoengineering' (McLellan *et al.* 2012, p. 1). Reviewing the burgeoning geoengineering economics literature, Hansson (2014) points out that the outcomes of these cost–benefit calculations can turn on discount rates that are themselves dependent on prior assumptions about the ethics of burdening future generations and whether geoengineering would add to or relieve the burden.

Geoengineering is only 'cheap and technically easy' (Keith 2013) if we make certain assumptions about what the technology is; if we don't question what it would mean for it to be 'effective'; if we choose to do some sums and not others; if we internalise some costs and externalise others; and if we assume that some things are easy, such as technological innovation, and other things are hard, such as political action. The economics of geoengineering are 'incredible' only in the sense that they are hard to believe as currently calculated, given what history tells us about the costs and benefits of complex sociotechnical systems.

Nuclear power requires extraordinary infrastructure in order to turn a powerful atomic reaction into a functioning sociotechnical system. Once the costs of making nuclear power stations reliable, secure and safe are taken into account, the technology looks expensive. It may have other benefits, albeit contested ones, but its direct costs are no lower than those of other forms of energy (see, for various assessments, RAEng 2004; Deutch *et al.* 2009; Cohen and McKillop 2012). And some would argue that many of the largest externalities, from proliferation threats and serious accidents to decommissioning, clean-up and waste disposal, are still not accounted for. With geoengineering, comparisons with civil nuclear power are less common than those with nuclear weapons. Geoengineering is seen as cheap in the way that building an atomic bomb is cheap (notwithstanding the multibillion dollar budget of the Manhattan Project).

The counterfactual costs of dramatically mitigating climate change or not doing anything have been more thoroughly estimated and challenged, but they are nonetheless still problematic. Economist William Nordhaus has used his dynamic integrated climate–economy (DICE) model to argue that geoengineering offers the potential for 'costless mitigation of climate change' (W. D. Nordhaus 1992, p. 1317). Nordhaus was one of the first economists to model climate change and a prominent critic of Nicholas Stern's review on the economics of climate change, commissioned by the UK government (W. D. Nordhaus 2007). Central to such disagreements are the discount rates – converting future costs into present-day ones – assumed by various economic models. Once different climate sensitivities and estimates of costs from extreme weather are included, the calculated 'social cost of carbon' implied can vary by an order of magnitude. According to one critical review, the 'models can be used to obtain almost any result one desires' (Pindyck 2013, p. 5). In searching for objective and politically persuasive information for climate policy, questions of what sort of world we would like to live in and the value we place on future generations are excised. The reduction of climate change to an economic problem, as opposed to just an engineering one, also allows a new angle for those such as Bjorn Lomborg (2010) who would suggest that if cheapness were the main criterion, other options including geoengineering should be considered.

Some economic research has begun to point out the uncertainties involved in making such calculations and the dangers of false certainty (e.g. Goes *et al.* 2011). But this research still sits in a tradition of cost–benefit analysis that assumes the possibility and value of balancing options on a single economic scale. Geoengineering would not be about weighing up costs or risks, even if we could perfectly calculate these. It would involve a radical change of direction, with radically new uncertainties and new relationships between society and the planet. Geoengineering will not solve climate change, which means that talk of remediation is disingenuous. Climate science and early geoengineering model results tell us that we cannot bring back past climates (e.g. Kravitz *et al.* 2013). And the history, philosophy and sociology of technology tell us that complex technologies take us down new paths from which we cannot retrace our steps (e.g. Latour and Venn 2002; Allenby 2013). Geoengineering would make the climate, not fix it. It is therefore more honest to talk about climate steering (Hale 2013, p. 201) or 'climate management' (Michaelson 2013). But this makes cost–benefit analysis an impotent tool for assessment, which risks justifying technologies that are unformed and unpredictable.

Wargames

Although the geoengineering bandwagon is still under construction, researchers from a wide range of disciplines are already jumping aboard. Geoengineering research has been able to accommodate insights and research questions from law, ethics, international relations, economics, environmental sociology, anthropology, media studies, science and technology studies and more besides. According to one scientometric analysis (Linnér and Wibeck, forthcoming), 2013 saw the number of published geoengineering papers outside the natural sciences surpass the number of scientific papers.

The vast bulk of the social science literature on geoengineering takes the scientific proposals at face value, albeit leaving behind the contingencies that the scientific researchers express at close quarters. Much of this work starts from the (flawed) assumption that geoengineering is so cheap that, in the words of David Victor *et al.* (2009, p. 71), 'a single country could deploy geoengineering systems from its own territory without consulting the rest of the planet'. Victor also imagines the possibility of a 'Greenfinger' – a billionaire Bond-villain type armed with a geoengineering technology (Victor 2008). According to Victor *et al.* (2009, p. 75), 'because the option exists, and might be used, it would be dangerous for scientists and policymakers to ignore it'.

Josh Horton (2011), without questioning the low-cost assumption, argues that this fixation on unilateralism is unrealistic and counterproductive. The technology as currently imagined would favour multilateralism. Nevertheless, such scenarios have been used to justify wargames in economics, game theory and international relations. Following Crutzen's lead, much of this research connects the use of geoengineering to a diagnosed failure of multilateral agreements to mitigate climate change. Some game theorists have concluded that geoengineering could solve the collective action problem of climate change by forcing nations to work together (Millard-Ball 2011). One group used a 'global thermostat setting game' to reach the same result (Ricke *et al.* 2013). Social scientists have collaborated with geoengineering scientists to run integrated assessment models on various components of the Earth system, such as food supply. Some of these studies have produced optimistic findings, pointing to, for example, an increased food supply in a geoengineered world (e.g. Pongratz *et al.* 2012). As I will explain in Chapter 7, there is little consideration in this research of the broader experimentality of geoengineering, which might suggest that dependence on such a technology would tend to increase instabilities in various ways, some of which we can anticipate and some of which will be unpredictable.

There has been no shortage of research connecting such considerations to matters of cost. One study modelled the costs and effects of climate change options and concluded, with a surprising degree of certainty, that the costs of climate change could be cut by something 'in the order of 10%' if SRM were granted additional research money (Moreno-Cruz and Keith 2012, p. 431).

This line of research presupposes that the uncertainties of geoengineering, even though all admit they extend well beyond the merely technical, can nevertheless be tamed through analysis and research. It is merely a matter of modelling the social and political alongside the scientific. There is little consideration that uncertainties and areas of ignorance might be so profound and intractable as to justify various other precautions. Indeed, there has been much discussion of how there is no clear precautionary approach to geoengineering (see Hartzell-Nicholls 2012 for a discussion), because it represents a 'risk/risk trade-off' (e.g. Barrett 2008, p. 52).

Following the refrain, clarified by the Royal Society, that 'the acceptability of geoengineering will be determined as much by social, legal and political issues as by scientific and technical factors' (Royal Society 2009, p. ix), researchers have begun to study pubic opinion of geoengineering to assess the 'barriers to public acceptance' (Jackson and Salzman 2010). Some of this work has taken the form of qualitative, open-ended public dialogue, in which the framing of geoengineering is not presupposed and researchers are careful not to reinforce particular frames (e.g. Pidgeon *et al.* 2012; Corner *et al.* 2013; Macnaghten and Szerszynski 2013). However, there is a growing research agenda that seeks to quantify public opinion about 'geoengineering' without challenging what the technology is or what it might mean (e.g. Mercer *et al.* 2011; M. J. Wright *et al.* 2014). Scheer and Renn (2014) review much of this literature and, though they don't question the way that geoengineering is understood by scientists or presented to members of the public, they do acknowledge that the uncertainties surrounding the technology and low public awareness mean that public opinions, forced in this way, may be better understood as 'pseudo-opinion'. As I will discuss in the next chapter, much of the impetus for this research is the supposed moral hazard that geoengineering is anticipated to introduce. For some scientists, this has been an invitation to treat public opinion and behaviour as a set of empirical questions, side-lining a more important discussion of the politics of geoengineering research.

Governance requires anticipation. We might therefore conclude that the types of multidisciplinary research that are springing up around geoengineering are therefore valuable. The surfacing of new, unintended possibilities has certainly been an important part of the move away from an unequivocal techno-optimism towards a mood of ambivalence towards geoengineering. But such research could be characterised as speculative ethics, and we should therefore follow Alfred Nordmann's (2007) critique described in the previous chapter. He argues that by fixating on the ethical, social, political, legal or economic implications of a particular technological future, we leapfrog an important set of discussions about governance in the present. Speculative ethics also represents an unconscious disavowal of responsibility. Scientists can convince themselves that a technology is sufficiently Promethean that someone will do it somehow, somewhere, and so they prevent themselves from having to think about whether they are making the technological future more likely. Similar arguments have been used to justify the acceleration of all manner of scientific investigations, from nuclear weapons research to the more recent case of the mutant influenza virus studies. Even though few geoengineering researchers wish to see a geoengineered world, there is an important dynamic of research in which geoengineering is being made thinkable across multiple disciplines.

Mapping geoengineering research

A scientometric analysis of geoengineering research, using data on published articles in journals, suggests that concerted scientific interest in geoengineering is a relatively recent phenomenon. As can be seen from the graph below (Figure 3.1), Paul Crutzen would seem to have either created or ridden a wave of

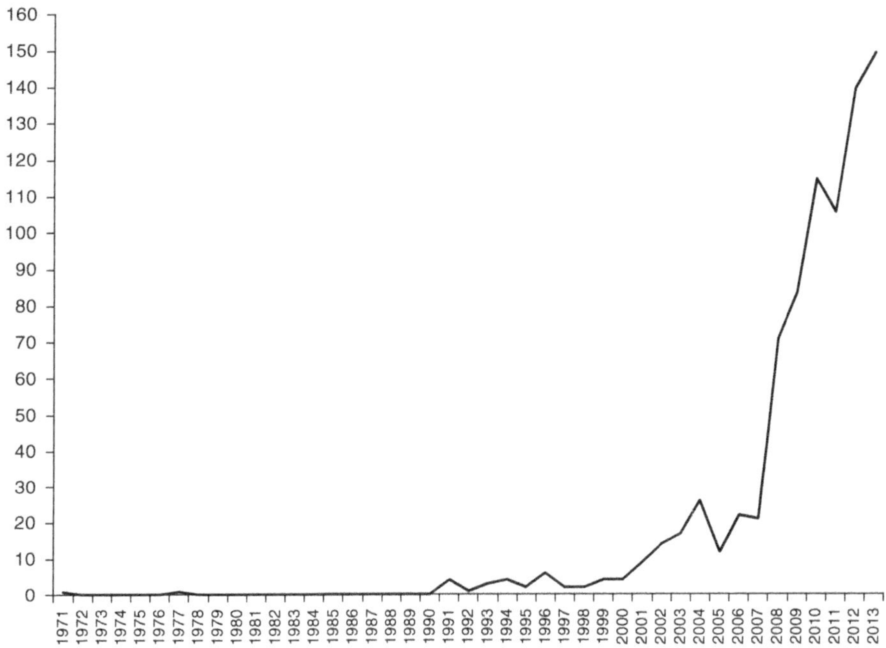

Figure 3.1 Number of geoengineering publications in the Web of Science database per year.

Source: Based on data from Oldham *et al.* (2014).[9]

scientific attention in 2006, and the same is true of the Royal Society in 2009. In interpreting the hockey-stick shape of this graph, however, we should remember that the numbers of publications remain low relative to emerging technology areas such as nanotechnology. Geoengineering research is growing, but it is not (yet) a large field of scientific research.[8] Much of the recent growth of geoengineering research has come from climate scientists. Very few articles are being published in engineering journals (Oldham *et al.* 2014).

The small world of geoengineering research has been called a 'geo-clique' by Eli Kintisch (2010), a claim repeated by Clive Hamilton (2011). The number of researchers is increasing with the number of publications, but a snapshot of the collaborations between authors suggests that there is indeed a relatively small group of researchers who are publishing large numbers of papers, indicated by the size of the circle in Figure 3.2. The author networks are shown in more detail in Figure 3.3.

Unthinking the unthinkable

The technologies imagined for geoengineering invite comparisons with nuclear weapons, both from scientists (e.g. Keith 2013) and in public-engagement discussions (e.g. Macnaghten and Szerszynski 2013). They are seen as sharing a Cold

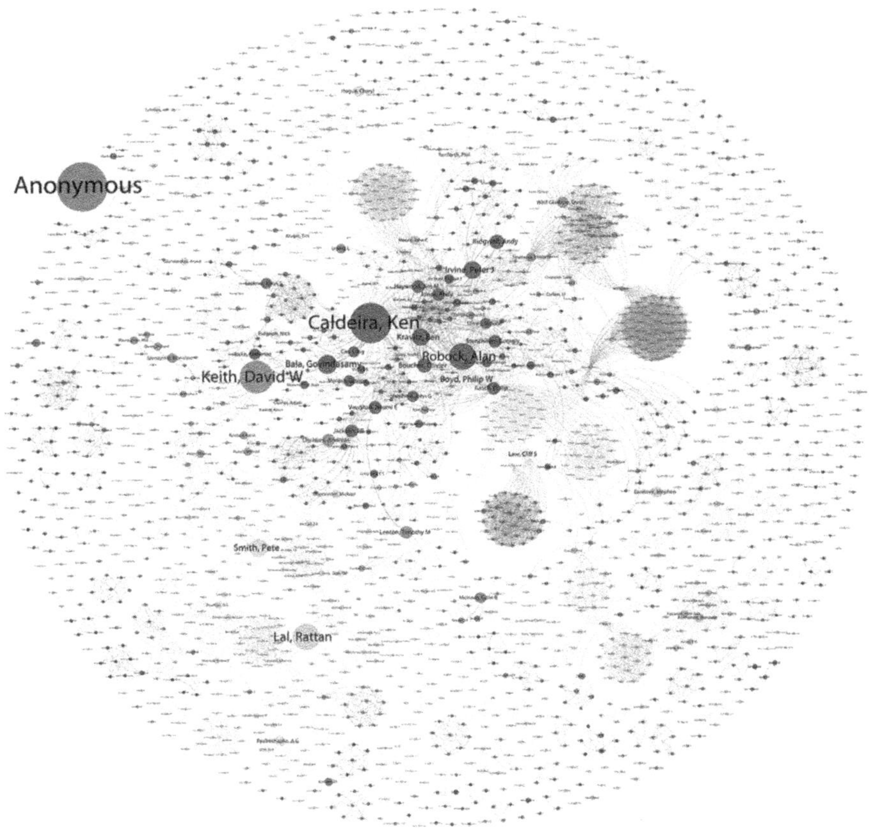

Figure 3.2 Author networks in geoengineering research.

Source: Based on data from Oldham *et al.* (2014).

War heritage, and they are seen as similarly potent. Nuclear weapons are the archetypical disruptive technology. They contain extraordinary potential power and therefore demand extraordinary control. With nuclear weapons, it seems unarguable that the things in themselves are dangerous. But social research suggests that their potency also depends on the tacit knowledge of nuclear-weapons researchers and the infrastructure that surrounds them. Donald Mackenzie and Graham Spinardi take issue with the Promethean narrative of nuclear weaponry, captured in a quote they take from the Harvard Nuclear Study Group: 'The discovery of nuclear weapons, like the discovery of fire itself, lies behind us on the trajectory of history. It cannot be undone' (quoted in MacKenzie and Spinardi 1995). Through looking at the tacit knowledge involved in the design, testing and maintenance of the technology, MacKenzie and Spinardi conclude that

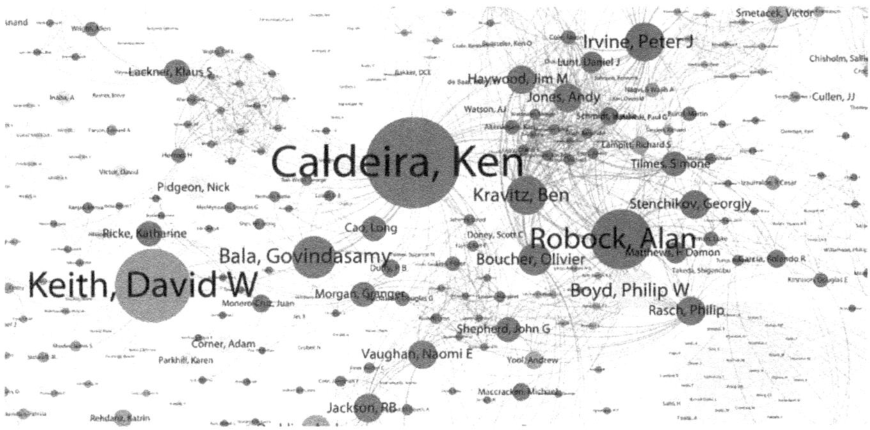

Figure 3.3 Detail of author networks.

Source: Based on data from Oldham *et al.* (2014).

nuclear weapons could in one respect be *uninvented* if generations of researchers were not around to keep this knowledge alive.

Nuclear weapons are substantially more complicated than an equation dictating the conversion of small amounts of matter into vast amounts of energy. In much the same way, the complications of geoengineering would stretch far beyond a fact about the climatic consequences of volcanic eruptions, and yet this complexity gets forgotten in the conventional story of geoengineering. Unlike nanotechnology or synthetic biology, geoengineering is an 'emerging technology' with no strong technoscientific basis. We are no more capable of geoengineering the climate than we were in the 1960s. Indeed, as climate science unfolds, scientists find themselves in some respects more uncertain about the possibilities of an engineered climate. We need therefore to think about the emergence of geoengineering in new ways – as an idea that has become 'thinkable'.

The usual history of geoengineering, such as it exists, is a story of potent, troublesome technologies, segregated from mainstream science by a fragile taboo that made it temporarily unthinkable. Clearly this does not mean it was logically impossible to think about geoengineering, but rather that its discussion was considered impolite, politically incorrect or illegitimate. Stephen Gardiner has discussed Bernard Williams's description of the unthinkable as the idea that 'entertaining certain alternatives, regarding them indeed as alternatives, is itself something that [someone] regards as dishonourable or morally absurd' (Gardiner 2011, p. 383). Such boundaries and taboos are protected by a familiar range of social and political tools, including dismissal, humour or intentional ignorance, from inside and outside the scientific community. For some, there is a presumptive argument from environmental ethics that geoengineering should remain unethical (Preston 2011). But taboos are only effective if they are observed nearly universally within a shared culture. The geoengineering taboo, on closer inspection, looks less like an opaque shroud than a wispy fog. Scientists getting close to geoengineering, once regarded as unthinkable, have found it eminently thinkable. Scientists, including most prominently Ken Caldeira, describe how they tried, using their models, to rule out the ideas of would-be geoengineers such as Lowell Wood but were taken by surprise by results that suggested it would be more effective and less dangerous than assumed. We should expect such experiments to generate surprises and challenge taboos. A 'see no evil, hear no evil' approach to geoengineering among the scientific community was always likely to be worse than useless, particularly if discussions were continuing behind closed doors (see Victor 2008; Preston 2011).

However, geoengineering should not be seen as a radical departure from research on climate change or Earth systems. We should instead look for the connections and continuities in order to recognise the possibility of alternative framings. Earth systems science has, over the last couple of decades, made the global climate and the interdependencies of humans and the environment scientifically thinkable (Lövbrand *et al.* 2009), in the same way that financial economics and accounting have made the global economy thinkable. The machinery of satellites, globally collaborative science and increasingly sophisticated computer models has built science's understanding of present and future climates. This has pushed many researchers to talk in terms humanity's responsibility for the climate, sometimes expressed as a form of experiment in which we are increasingly self-conscious participants. Ideas of the Anthropocene and planetary boundaries, though clothed in scientific language and fought about in scientific terms, are as much political responses to a perceived lack of political action to deal with climate change. We are witnessing the continued advancement of technoscientific prediction and control over a debate that is unavoidably also political. Against this backdrop the arrival, or rather encroachment, of geoengineering looks less surprising. In the next chapter, I explain how the Royal Society's assessment of geoengineering also contributed to the construction of the issue and its imagined technologies.

Notes

1 With thanks to Roger Pielke, Jr for directing me to this.
2 With thanks to Chris Stokes for pointing me to this one.
3 A description used by, among others, *Washington Post* columnist Robert Samuelson (2006) and Rex Tillerson from Exxon (Daily 2012).
4 Translated from the 1974 Russian version by the American Geophysical Union: Budyko, M. I. (1977). *Climatic changes*. AGU Special Publication 10. Baltimore: Waverly Press.
5 This example is borrowed from Alan Robock.
6 Raymond Pierrehumbert, quoted in Rotman (2013).
7 Ken Caldeira, email posted to the Geoengineering Google Group, 2 October 2013.
8 The methodological detail behind the generation and analysis of this dataset is given in Oldham *et al.* (2014). Generating this data requires many choices about what to look

for, what to include and why. Justifications are provided in the paper, but the dataset is by no means the final word on the matter.

9 *Web of Science* database: http://thomsonreuters.com/thomson-reuters-web-of-science/.

References

Allen, M. (2009). Commentary. *Nature Reports: Climate Change*, 3(10), n.pp. Available online at http://www.nature.com/climate/2009/0910/full/climate.2009.93.html (accessed 19 June 2014).

Allenby, B. (2013). An Earth systems engineering perspective on geoengineering. *Proceedings of the ICE: Engineering Sustainability*, 166(5), 220–228.

Angel, R. (2006). Feasibility of cooling the Earth with a cloud of small spacecraft near the inner Lagrange point (L1). *Proceedings of the National Academy of Sciences*, 103(46), 17184–17189.

Asilomar Scientific Organizing Committee (2010). *The Asilomar Conference recommendations on principles for research into climate engineering techniques*. Conference Report. Washington, DC: Climate Institute. Available online at http://climate.org/PDF/AsilomarConferenceReport.pdf (accessed 20 June 2012).

Badiou, A. (2008). Live Badiou – interview with Alain Badiou, Paris, December 2007. In O. Feltham (ed.), *Alain Badiou: live theory* (pp. 136–139). London: Continuum.

Barrett, S. (2008). The incredible economics of geoengineering. *Environmental & Resource Economics*, 39(1), 45–54.

Baskin, J. (2014). *The ideology of the Anthropocene?* Research Paper No. 3. Victoria: Melbourne Sustainable Society Institute, University of Melbourne.

Beck, U. (2010). Climate for change, or how to create a green modernity? *Theory, Culture & Society*, 27(2/3), 254–266.

Berkner, L. V. (1962). Geoscience and geoengineering. *Proceedings of the IRE*, 50(11), 2180–2182.

Bernal, J. D. (1939). *The social function of science*, 2010 edn. London: Faber & Faber.

Borenstein, S. (2007). Some scientists eye odd climate fixes. *Washington Post*, 18 March. Available online at http://www.washingtonpost.com/wp-dyn/content/article/2007/03/18/AR2007031800643.html (accessed 1 June 2014).

Borup, M., Brown, N., Konrad, K., and van Lente, H. (2006). The sociology of expectations in science and technology. *Technology Analysis & Strategic Management*, 18(3/4), 285–298.

Brand, S. (2010). *Whole Earth discipline: why dense cities, nuclear power, transgenic crops, restored wildlands, and geoengineering are necessary*. New York: Penguin.

Bray, D., and von Storch, H. (2009). 'Prediction' or 'projection'? The nomenclature of climate science. *Science Communication*, 30(4), 534–543.

Briffa, K. R., Jones, P. D., Schweingruber, F. H., and Osborn, T. J. (1998). Influence of volcanic eruptions on Northern Hemisphere summer temperature over the past 600 years. *Nature*, 393(6684), 450–455.

Budyko, M. I. (1974). *Izmeniya klimata*. Leningrad: Gidrometeoizdat.

Chakrabarty, D. (2009). The climate of history: four theses. *Critical Inquiry*, 35(2), 197–222.

Cicerone, R. J. (2006). Geoengineering: encouraging research and overseeing implementation. *Climatic Change*, 77(3), 221–226.

Cohen, M., and McKillop, A. (2012). *The Doomsday machine: the high price of nuclear energy, the world's most dangerous fuel*. New York: Palgrave Macmillan.

Corner, A., Parkhill, K., Pidgeon, N., and Vaughan, N. E. (2013). Messing with Nature? Exploring public perceptions of geoengineering in the UK. *Global Environmental Change*, 23(5), 938–947.

Crutzen, P. J. (2002). Geology of mankind. *Nature*, 415(6867), 23.

—— (2006). Albedo enhancement by stratospheric sulfur injections: a contribution to resolve a policy dilemma? *Climatic Change*, 77(3), 211–220.

Crutzen, P. J., and Steffen, W. (2003). How long have we been in the Anthropocene era? *Climatic Change*, 61(3), 251–257.

Crutzen, P. J., and Stoermer, E. F. (2000). The 'Anthropocene'. *Global Change Newsletter*, 41, 17–18.

Daily, M. (2012). Exxon CEO calls climate change engineering problem. Reuters, 27 June. Available online at http://in.reuters.com/article/2012/06/27/exxon-climate-idINL2E8HRA8K20120627 (accessed 29 May 2014).

Dando, M., Pearson, G., Rózsa, L., Robinson, J. P., and Wheelis, M. (2006). Analysis and implications. In M. Wheelis, L. Rózsa and M. Danda, *Deadly cultures: biological weapons since 1945* (pp. 355–374). Cambridge, MA: Harvard University Press.

Davidson, P., Burgoyne, C., Hunt, H., and Causier, M. (2012). Lifting options for stratospheric aerosol geoengineering: advantages of tethered balloon systems. *Philosophical Transactions of the Royal Society A: Mathematical, Physical & Engineering Sciences*, 370(1974), 4263–4300.

Deutch, J. M., Forsberg, C. W., Kadak, A. C., Kizimi, M. S., Moniz, P., *et al.* (2009). *Update of the MIT 2003 Future of nuclear power*. MIT Energy Initiative. Cambridge, MA: Massachusetts Institute of Technology.

Dilling, L., and Hauser, R. (2013). Governing geoengineering research: why, when and how? *Climatic Change*, 121(3), 553–565.

Douglas, M. (1992). *Risk and blame: essays in cultural theory*. London: Routledge.

Edwards, P. N. (2001). Representing the global atmosphere: computer models, data, and knowledge about climate change. In C. A. Miller and P. N. Edwards (eds), *Changing the atmosphere: expert knowledge and environmental governance* (pp. 31–36). Cambridge, MA: MIT Press.

——— (2010). *A vast machine: computer models, climate data, and the politics of global warming*. Cambridge, MA: MIT Press.

ENS (2009). A Copenhagen climate accord, not nothing, but not enough. *Environment News Service*, 19 December. Available online at http://ens-newswire.com/2009/12/19/a-copenhagen-climate-accord-not-nothing-but-not-enough/, (accessed 29 May 2014).

ETC Group (2010). *Sign on! Open letter opposing Asilomar Geoengineering Conference*. Ottawa: ETC Group. Available online at http://www.etcgroup.org/content/sign-open-letter-opposing-asilomar-geoengineering-conference (accessed 10 June 2014).

Feichter, J. (2011). Shaping reality with algorithms: the Earth system. In G. Gramelsberger (ed.), *From science to computational sciences: studies in the history of computing and its influence on today's sciences* (pp. 209–218). Zurich: Diaphanes.

Fleming, J. (2010). *Fixing the sky: the checkered history of weather and climate control*. New York: Columbia University Press.

——— (2012). Will geo-engineering bring security and peace? What does history tell us? *Sicherheit und Frieden (Security and Peace)*, 30(4), 200–204.

Galison, P., and Bernstein, B. J. (1989). In any light: scientists and the decision to build the superbomb, 1952–1954. *Historical Studies in the Physical and Biological Sciences*, 19(2), 267–347.

Gardiner, S. M. (2011). *A perfect moral storm: the ethical tragedy of climate change*. Oxford: Oxford University Press.

Goes, M., Tuana, N., and Keller, K. (2011). The economics (or lack thereof) of aerosol geoengineering. *Climatic Change*, 109(3/4), 719–744.

Goodell, J. (2010). A hard look at the perils and potential of geoengineering. *Yale Environment 360*, 1 April. Available online at http://e360.yale.edu/feature/a_hard_look_at_the_perils_and_potential_of_geoengineering/2260/ (accessed 11 June 2014).

Gore, A. (2009). *Our choice: a plan to solve the climate crisis*. Emmaus, PA: Rodale Books.

Gramelsberger, G., and Feichter, J. (2011). Modelling the climate system: an overview. In G. Gramelsberger and J. Feichter (eds), *Climate change and policy: the calculability of climate change and the challenge of uncertainty* (pp. 9–90). Heidelberg: Springer.

Grundmann, R. (2007). Climate change and knowledge politics. *Environmental Politics*, 16(3), 414–432.

Guo, S., Rose, W. I., Bluth, G. J. S., and Watson, M. (2004). Particles in the great Pinatubo volcanic cloud of June 1991: the role of ice. *Geochemistry, Geophysics, Geosystems*, 5(5), 1–35.

Hale, B. (2013). Remediation vs. steering: an act-description approach to approving and funding geoengineering research. In J. Basl and R. L. Sandler (eds), *Designer biology: the ethics of intensively engineering biological and ecological systems* (pp. 197–217). New York: Lexington Books.

Hall, A. (2012). *Risk, blame, and expertise: The Meteorological Office and extreme weather in post-war Britain*. PhD thesis, University of Manchester, Manchester.

Hamblin, J. D. (2013). *Arming Mother Nature: the birth of catastrophic environmentalism*. New York: Oxford University Press.

Hamilton, C. (2011). The clique that is trying to frame the global geoengineering debate. *The Guardian*, 5 December. Available online at http://www.theguardian.com/environment/2011/dec/05/clique-geoengineering-debate (accessed 29 July 2014).

——— (2013). *Earthmasters: the dawn of the age of climate engineering*. New Haven, CT: Yale University Press.

Hansson, A. (2014). Ambivalence in calculating the future: the case of re-engineering the world. *Journal of Integrative Environmental Sciences*, 11(2), 125–142.

Harper, K. C. (2006). Meteorology's struggle for professional recognition in the USA (1900–1950). *Annals of Science*, 63(2), 179–199.

——— (2008). *Weather by the numbers: the genesis of modern meteorology*. Cambridge, MA: MIT Press.

——— (2009). *Weather and climate: decade by decade*. New York: Infobase Publishing.

Hartzell-Nichols, L. (2012). Precaution and solar radiation management. *Ethics, Policy & Environment*, 15(2), 158–171.

Hastrup, K. (2013). Anticipating Nature: the productive uncertainty of climate models. In K. Hastrup and M. Skrydstrup (eds), *Routledge studies in anthropology*, Vol. 8: *The social life of climate change models: anticipating Nature* (pp. 1–29). London: Routledge.

Hilgartner, S. (2007). Overflow and containment in the aftermath of disaster. *Social Studies of Science*, 37(1), 153–158.

Horton, J. B. (2011). Geoengineering and the myth of unilateralism: pressures and prospects for international cooperation. *Stanford Journal of Law, Science & Policy*, 4(1), 56–69.

Horton, Z. (2014). Collapsing scale: nanotechnology and geoengineering as speculative media. In K. Konrad, C. Coenen, A. Dijkstra, C. Milburn and H. van Lente, *Studies of new and emerging technologies*, Vol. 4: *Shaping emerging technologies: governance, innovation, discourse* (pp. 203–218). Amsterdam: IOS Press.

Hulme, M., O'Neill, S. J., and Dessai, S. (2011). Is weather event attribution necessary for adaptation funding? *Science*, 334(6057), 764–765.
IPCC (2001). Technical summary. In J. T. Houghton, Y. Ding, D. J. Griggs, M. Noguer, P. J. van der Linden *et al.*, *Climate change 2001: the scientific basis* (pp. 17–786). Contribution of Working Group I to the Third Assessment Report of the Intergovernmental Panel on Climate Change. New York: Cambridge University Press.
Jackson, R. B., and Salzman, J. (2010). Pursuing geoengineering for atmospheric restoration. *Issues in Science and Technology*, 26(4), 67–76.
Jasanoff, S., and Martello, M. L. (eds) (2004). *Earthly politics: local and global in environmental governance*. Cambridge, MA, and London: MIT press.
Keith, D. W. (2000). Geoengineering the climate: history and prospect. *Annual Review of Energy and the Environment*, 25, 245–284.
——— (2010). Engineering the planet. In S. H. Schneider, A. Rosencranz, M. D. Mastrandrea, and K. Kuntz-Duriseti (eds), *Climate change science and policy* (pp. 494–501). Washington, DC: Island Press.
——— (2013). *A case for climate engineering*. Cambridge, MA: MIT Press.
Keith, D. W., and Dowlatabadi, H. (1992). A serious look at geoengineering. *Eos, Transactions of the American Geophysical Union*, 73(27), 289–293.
Kintisch, E. (2010). *Hack the planet: science's best hope – or worst nightmare – for averting climate catastrophe*. Hoboken, NJ: Wiley.
Kravitz, B., Caldeira, K., Boucher, O., Robock, A., Rasch, P. J., *et al.* (2013). Climate model response from the Geoengineering Model Intercomparison Project (GeoMIP). *Journal of Geophysical Research: Atmospheres*, 118(15), 8320–8332.
Krimsky, S. (2005). From Asilomar to industrial biotechnology: risks, reductionism and regulation. *Science as Culture*, 14(4), 309–323.
Latham, J. (2002). Amelioration of global warming by controlled enhancement of the albedo and longevity of low-level maritime clouds. *Atmospheric Science Letters*, 3(2/3/4), 52–58.
Latour, B. (1999). *Pandora's hope: essays on the reality of science studies*. Cambridge, MA: Harvard University Press.
——— (2013). Inside the 'planetary boundaries': Gaia's estate. Gifford Lecture 6 in the series *Facing Gaia: A New Enquiry into Natural Religion*, 28 February. University of Edinburgh, Edinburgh.
Latour, B., and Venn, C. (2002). Morality and technology: the end of the means. *Theory, Culture & Society*, 19(5/6), 247–260.
Lawrence, M. G. (2006). The geoengineering dilemma: to speak or not to speak. *Climatic Change*, 77(3), 245–248.
Lenhard, J. (2007). Computer simulation: the cooperation between experimenting and modeling. *Philosophy of Science*, 74(2), 176–194.
Levitt, S. D., and Dubner, S. J. (2005). *Freakonomics: a rogue economist explores the hidden side of everything*. New York: HarperCollins.
——— (2010). *Superfreakonomics: global cooling, patriotic prostitutes and why suicide bombers should buy life insurance*. London: Penguin UK.
Linnér, B-O., and Wibeck, V. (Forthcoming). Climate engineering as emerging technologies – a review of the literature. *Wiley Interdisciplinary Reviews: Climate Change*. (Submitted.)
Lomborg, B. (ed.) (2010). *Smart solutions to climate change: comparing costs and benefits*. New York: Cambridge University Press.
Lövbrand, E., Stripple, J., and Wiman, B. (2009). Earth system governmentality: reflections on science in the Anthropocene. *Global Environmental Change*, 19(1), 7–13.

Lovelock, J. (2009). *The vanishing face of Gaia: a final warning*. London: Allen Lane.

MacKenzie, D., and Spinardi, G. (1995). Tacit knowledge, weapons design, and the uninvention of nuclear weapons. *American Journal of Sociology*, 101(1), 44–99.

MacKerron, G. (2014). *Costs and economics of geoengineering*. CGG Working Paper No. 13. Oxford: Climate Geoengineering Governance, Institute for Science, Innovation and Society. Available online at http://www.geoengineering-governance-research.org/perch/resources/workingpaper13mackerroncostsandeconomicsofgeoengineering.pdf (accessed 18 July 2014).

Macnaghten, P., and Szerszynski, B. (2013). Living the global social experiment: an analysis of public discourse on solar radiation management and its implications for governance. *Global Environmental Change*, 23(2), 465–474.

Marchetti, C. (1977). On geoengineering and the CO_2 problem. *Climatic Change*, 1(1), 59–68.

Markusson, N., Ginn, F., Singh Ghaleigh, N., and Scott, V. (2013). 'In case of emergency press here': framing geoengineering as a response to dangerous climate change. *Wiley Interdisciplinary Reviews: Climate Change*, 5(2), 281–290.

Marland, G. (1996). Could we/should we engineer the Earth's climate? *Climatic Change*, 33(3), 275–278.

Mathews, A. (2011). *Geoengineering as terraformation: a constitutional convention on the future*. Presented at the 110th Annual Meeting of the American Anthropological Association, Montréal, 16–20 November 2011.

McClellan, J., Keith, D. W., and Apt, J. (2012). Cost analysis of stratospheric albedo modification delivery systems. *Environmental Research Letters*, 7(3), 034019.

McClellan, J., Sisco, J., Suarez, B., and Keogh, G. (2010). *Geoengineering cost analysis*. Final Report. Cambridge, MA: Aurora Flight Sciences Corporation.

McInnes, C. (2013). Time to embrace geoengineering – beyond planetary boundaries. *Breakthrough Journal*, 27 June. Available online at http://thebreakthrough.org/index.php/programs/energy-and-climate/time-to-embrace-geoengineering (accessed 11 March 2014).

McKibben, B. (1989). *The end of Nature*. New York: Random House.

Mercer, A. M., Keith, D. W., and Sharp, J. D. (2011). Public understanding of solar radiation management. *Environmental Research Letters*, 6(4), 044006.

Michaelson, J. (2013). Geoengineering and climate management: from marginality to inevitability. In W. C. G. Burns and A. L. Strauss (eds), *Climate change geoengineering: philosophical perspectives, legal issues and governance frameworks* (pp. 81–114). New York: Cambridge University Press.

Millard-Ball, A. (2011). The Tuvalu syndrome. *Climatic Change*, 110(3/4), 1047–1066.

Moreno-Cruz, J. B., and Keith, D. W. (2012). Climate policy under uncertainty: a case for solar geoengineering. *Climatic Change*, 121(3), 431–444.

Morgan, M. S. (2013). Nature's experiments and natural experiments in the social sciences. *Philosophy of the Social Sciences*, 43(3), 341–357.

Morton, O. (2007). Climate change: is this what it takes to save the world? *Nature*, 447(7141), 132–136.

NAS (1992). *Policy implications of greenhouse warming*. Washington, DC: National Academies Press.

Nelkin, D. (2001). Beyond risk: reporting about genetics in the post-Asilomar press. *Perspectives in Biology and Medicine*, 44(2), 199–207.

Nordhaus, T., and Shellenberger, M. (2007). *Break through: from the death of environmentalism to the politics of possibility*. New York: Houghton Mifflin.

Nordhaus, T., Shellenberger, M., and Blomqvist, L. (2012). *The planetary boundaries hypothesis: a review of the evidence*. Oakland, CA: Breakthrough Institute.

Nordhaus, W. D. (1992). An optimal transition path for controlling greenhouse gases. *Science*, 258(5086), 1315–1319.

——— (2007). A review of the 'Stern review on the economics of climate change'. *Journal of Economic Literature*, 45(Sept), 686–702.

Nowotny, H. (2007). *Knowledge production and its constraints: epistemic and societal considerations*. Presented at the Gulbenkian Foundation Conference, Is Science Nearing Its Limits? Lisbon, 25–26 October 2007. Available online at http://www.helga-nowotny.eu/downloads/helga_nowotny_b58.pdf (accessed 23 May 2014).

Nowotny, H., and Testa, G. (2010). *Naked genes: reinventing the human in the molecular age*. Cambridge, MA: MIT Press.

NSF (1966). *Weather and climate modification*. Report of the Special Commission on Weather Modification. Washington, DC: National Science Foundation.

Oldham, P., Szerszynski, B., Stilgoe, J., Brown, C., Eacott, B., and Yuille, A. (2014). Mapping the landscape of climate engineering. *Philosophical Transactions of the Royal Society A: Mathematical, Physical & Engineering Sciences*, 372(2031), 20140065.

Olson, R. L. (2012). Soft geoengineering: a gentler approach to addressing climate change. *Environment: Science and Policy for Sustainable Development*, 54(5), 29–39.

Oreskes, N., Shrader-Frechette, K., and Belitz, K. (1994). Verification, validation, and confirmation of numerical models in the Earth sciences. *Science*, 263(5147), 641–646.

Pidgeon, N., Corner, A., Parkhill, K., Spence, A., Butler, C., and Poortinga, W. (2012). Exploring early public responses to geoengineering. *Philosophical Transactions of the Royal Society A: Mathematical, Physical & Engineering Sciences*, 370(1974), 4176–4196.

Pielke, R., Jr (2013). Planetary boundaries as power grab. *Roger Pielke Jr.'s Blog*. Available online at http://rogerpielkejr.blogspot.co.uk/2013/04/planetary-boundries-as-power-grab.html (accessed 23 May 2014).

Pindyck, R. S. (2013). *Climate change policy: what do the models tell us?* No. w19244. Cambridge, MA: National Bureau of Economic Research.

Pongratz, J., Lobell, D. B., Cao, L., and Caldeira, K. (2012). Crop yields in a geoengineered climate. *Nature Climate Change*, 2(2), 101–105.

Preston, C. J. (2011). Re-thinking the unthinkable: environmental ethics and the presumptive argument against geoengineering. *Environmental Values*, 20(4), 457–479.

PSAC (1965). *Restoring the quality of our environment*. Washington, DC: Executive Office of the President, President's Science Advisory Committee.

RAEng (2004). *The cost of generating electricity*. London: Royal Academy of Engineering.

Randalls, S. (2010). Weather profits: weather derivatives and the commercialization of meteorology. *Social Studies of Science*, 40(5), 705–730.

Rayner, S. (2012). Uncomfortable knowledge: the social construction of ignorance in science and environmental policy discourses. *Economy and Society*, 41(1), 107–125.

Rayner, S., Heyward, C., Kruger, T., Pidgeon, N., Redgwell, C., and Savulescu, J. (2013). The Oxford Principles. *Climatic Change*, 121(3), 499–512.

Revkin, A. C. (2011). Confronting the 'Anthropocene'. *The Opinion Pages, Dot Earth* blog, 11 May. Available online at http://dotearth.blogs.nytimes.com/2011/05/11/confronting-the-anthropocene/?_php=true&_type=blogs&_r=0 (accessed 1 May 2014).

Ricke, K. L., Moreno-Cruz, J. B., and Caldeira, K. (2013). Strategic incentives for climate geoengineering coalitions to exclude broad participation. *Environmental Research Letters*, 8(1), 014021.

Robock, A. (2008). 20 reasons why geoengineering may be a bad idea. *Bulletin of the Atomic Scientists*, 64(2), 14–18, 59.

——— (2014). A case against climate engineering. *The Blog, Huffington Post*, 5 May. Available online at http://www.huffingtonpost.com/alan-robock/a-case-against-climate-engineering_b_5264200.html (accessed 30 May 2014).

Rockström, J., Steffen, W., Noone, K., Persson, Å., Chapin, F. S., *et al.* (2009). A safe operating space for humanity. *Nature*, 461(7263), 472–475.

Rogers, M. (1975). The Pandora's box Congress. *Rolling Stone*, 189(19 Jun), 37–40, 42, 74, 77–78, 82.

Rotman, D. (2013). A cheap and easy plan to stop global warming. *MIT Technology Review*, 8 February. Available online at http://www.technologyreview.com/featuredstory/511016/a-cheap-and-easy-plan-to-stop-global-warming/ (accessed 31 May 2014).

Royal Society (2009). *Geoengineering the climate: science, governance and uncertainty*. London: Royal Society.

Ruddiman, W. F. (2003). The anthropogenic greenhouse era began thousands of years ago. *Climatic Change*, 61(3), 261–293.

Samuelson, R. (2006). Global warming's real inconvenient truth. *Washington Post*, 5 July. Available online at http://www.washingtonpost.com/wp-dyn/content/article/2006/07/04/AR2006070400789.html (accessed 29 May 2014).

Sarewitz, D. (2004). How science makes environmental controversies worse. *Environmental Science & Policy*, 7(5), 385–403.

Sarewitz, D., and Pielke, R. A., Jr (2000). Breaking the global warming gridlock. *Atlantic Monthly*, 286(1), 55–64.

Schäfer, S., and Low, S. (2014). Asilomar moments: formative framings in recombinant DNA and solar climate engineering research. *Philosophical Transactions of the Royal Society A: Mathematical, Physical & Engineering Sciences*, 372(2031), 20140064.

Scheer, D., and Renn, O. (2014). Public perception of geoengineering and its consequences for public debate. *Climatic Change*, 125(3/4), 305–318.

Schelling, T. C. (1996). The economic diplomacy of geoengineering. *Climatic Change*, 33(3), 303–307.

Schlesinger, W. H. (2009). Planetary boundaries: thresholds risk prolonged degradation. Commentary. *Nature Reports: Climate Change*, 3(10). n.pp. Available online at http://www.nature.com/climate/2009/0910/full/climate.2009.93.html (accessed 19 June 2014).

Schneider, S. H. (1997). *Laboratory Earth: the planetary gamble we can't afford to lose*. New York: Basic Books.

——— (2008). Geoengineering: could we or should we make it work? *Philosophical Transactions of the Royal Society A: Mathematical, Physical & Engineering Sciences*, 366(1882), 3843–3862.

Schuurbiers, D. (2011). What happens in the lab: applying midstream modulation to enhance critical reflection in the laboratory. *Science and Engineering Ethics*, 17(4), 769–788.

Scott, K. N. (2013). International law in the Anthropocene: responding to the geoengineering challenge. *Michigan Journal of International Law*, 34(2), 309–358.

Shapin, S. (2002). Megaton man. *London Review of Books*, 28(8), 18–20.

Steffen, A. (2008). Planktos, geo-engineering and politics. *Worldchanging.com*, 14 February 2008. Available online at http://www.worldchanging.com/archives//007841.html (accessed 1 June 2014).

Steffen, W., Persson, A., Deutsch, L., Zalasiewiecz, J., Williams, M., *et al.* (2011a). The Anthropocene: from global change to planetary stewardship. *AMBIO*, 40(7), 739–761.

Steffen, W., Rockström, J., and Costanza, R. (2011b). How defining planetary boundaries can transform our approach to growth. *Solutions*, 2(3). Available online at http://www.thesolutionsjournal.com/node/935?page=35%2C0%2C0%2C0%2C0%2C0 (accessed 29 June 2014).

Steffen, W., Grinevald, J., Crutzen, P., and McNeill, J. (2012). The Anthropocene: conceptual and historical perspectives. *Philosophical Transactions of the Royal Society A: Mathematical, Physical & Engineering Sciences*, 369(1938), 842–867.

Stehr, N. (2013). An inconvenient democracy: knowledge and climate change. *Society*, 50(1), 55–60.

Steinberg, T. (2006). *Acts of God: the unnatural history of natural disaster in America*, 2nd edn. New York: Oxford University Press.

Stilgoe, J. (2007). The (co-)production of public uncertainty: UK scientific advice on mobile phone health risks. *Public Understanding of Science*, 16(1), 45–61.

Stilgoe, J., Watson, M., and Kuo, K. (2013). Public engagement with biotechnologies offers lessons for the governance of geoengineering research and beyond. *PLoS Biology*, 11(11), e1001707.

Teller, E., Hyde, R., and Wood, L. (2002). *Active climate stabilization: practical physics-based approaches to prevention of climate change*. Livermore, CA: University of California, Lawrence Livermore National Laboratory.

Teller, E., Wood, L., and Hyde, R. (1997). *Global warming and ice ages: I. Prospects for physics based modulation of global change*. Livermore, CA: University of California, Lawrence Livermore National Laboratory.

Trenberth, K. E., and Dai, A. (2007). Effects of Mount Pinatubo volcanic eruption on the hydrological cycle as an analog of geoengineering. *Geophysical Research Letters*, 34(15), L15702.

Urry, J. (2011). *Climate change and society*. Cambridge: Polity Press.

Varenne, F. (2001). What does a computer simulation prove? The case of plant modeling at CIRAD (France). In N. Giambiasi and C. Frydman (eds), *Simulation in industry, proceedings of the 13th European Simulation Symposium, Marseille, France, 18–20 October 2001* (pp. 549–554). Ghent, Belgium: SCS Europe Bvba.

Victor, D. (2008). On the regulation of geoengineering. *Oxford Review of Economic Policy*, 24(2), 322–336.

Victor, D. G., Morgan, M. G., Apt, J., Steinbruner, J., and Ricke, K. (2009). The geoengineering option: a last resort against global warming? *Foreign Affairs*, 88(2), 64–76.

——— (2013). The truth about geoengineering: science fiction and science fact. *Foreign Affairs*, 92(2), Postscript.

Weart, S. R. (2008). *The discovery of global warming*. Cambridge, MA: Harvard University Press.

Wigley, T. M. (2006). A combined mitigation/geoengineering approach to climate stabilization. *Science*, 314(5798), 452–454.

Wood, G. D'A. (2014). *Tambora: the eruption that changed the world*. Princeton: Princeton University Press.

Wright, M. J., Teagle, D. A. H., and Feetham, P. M. (2014). A quantitative evaluation of the public response to climate engineering. *Nature Climate Change*, 4(2), 106–110.

Wright, S. (2001). Legitimating genetic engineering. *Perspectives in Biology and Medicine*, 44(2), pp. 235–247.

Wynne, B. (2010). Strange weather, again climate science as political art. *Theory, Culture & Society*, 27(2/3), 289–305.

4 Behind the scenes at the Royal Society

On Tuesday 1 September 2009, the morning after the UK's summer bank holiday, the stage was set for a bravura performance of scientific advice on geoengineering. The cast of speakers was mixed. John Beddington, the government chief scientific adviser, James Lovelock, the outspoken independent scientist behind the Gaia hypothesis, and Greenpeace's chief scientist Doug Parr joined members of the Royal Society Working Group on Geoengineering.[1] On the report's cover, underneath the title, *Geoengineering the Climate: Science, Governance and Uncertainty*, sat an image based on Watson and Lovelock's Daisyworld, a groundbreaking computer simulation from the 1980s that modelled the co-evolution of an environment and its organisms, showing a planetary system's response to sunlight reflection (Royal Society 2009b). Even with this varied cast, the Society, a veritable institution with a long history, managed to pull off a piece of coherent modern theatre.

Their event was introduced by Martin Rees. As well as being the Society's president, Lord Rees was at the time in possession of two other centuries-old titles: Master of Trinity College, Cambridge and Astronomer Royal. He sat at the very top of the British scientific establishment, but his approach was one of quiet activism rather than tub-thumping. For the launch of this report, Rees had a clear sense of the stage directions, but he had written his own lines. In the foreword to the report and in his introductory comments at the launch, he explained scientists' concern about climate change and the growing pressure from some quarters to consider a geoengineering plan B. The Royal Society had detected that the debate about geoengineering had become detached from good science and was rolling like a loose cannon, with the potential to destabilise the delicate climate negotiations that were due to take place later in 2009. Their report was an attempt to reintroduce some scientific rigour. In the years since its publication, the report remains the preeminent reference point for debates about geoengineering and its governance. It is therefore worthy of close examination in its own right. However, its appearance also raises the question of how a topic that had been considered unthinkable in polite scientific company only a few years before became worthy of consideration by the world's oldest science academy.[2]

An age of experiments

The Royal Society began as 'A Colledge for the Promoting of Physico-Mathematicall Experimentall Learning' in 1660. This 'invisible college', having received royal approval from Charles II, was renamed 'The Royal Society for Improving Natural Knowledge'. The idea that the experiments conducted and witnessed by Society Fellows should speak for themselves was enshrined in the motto: *Nullius in Verba* – 'Take no man's word for it'. According to Lisa Jardine, the Royal Society gained much of its early influence from the experimental talents of Robert Hooke, the Society's first curator of experiments (Jardine 2003), making him perhaps the first professional scientist, albeit in an era that predated the word 'scientist' (May 2005). Among his first contributions was, in 1663, a 'method for making the history of the weather' (Sprat 1958).

Hooke's aim was for the Royal Society 'to improve the knowledge of natural things, and all useful Arts, Manufactures, Mechanic practices, Engines and Inventions by Experiments (not meddling with divinity, Metaphysics, Morals, Politics, Grammar, Rhetoric or Logic)' (quoted in van den Daele 1977). The Society's interest in technology remained until the 1960s, when growing policy interest in engineering led to the formation of a separate institution, the Royal Academy of Engineering (RAEng) (Collins 2010). Its relationship with philosophy, the social sciences (which might in the seventeenth century have been lumped under 'Grammar') and politics were more complicated.

The Royal Society has always told stories about itself, and its history as part of its claim to political authority. Its self-mythologising began at an early age. Thomas Sprat wrote the first history of the Royal Society in 1667, seven years after its founding. This 'history' was more prospective than retrospective, acting instead as an official manifesto for the Society's engagements with the world. Sprat defined the time as an 'age of experiments' (although Steven Shapin [2011] notes that this was a wry nod to a much-mocked innovation of the time: William Petty's catamaran, christened *The Experiment*). Right from the start, however, the Royal Society was as ambitious in the social work of gaining political authority as in its experimental work. Its first policy report came in 1664, advising the Royal Navy on the state of Britain's forests following a massive shipbuilding programme, with the admirably concise title *Sylva*. (The length of the subtitle, *A Discourse of Forest-Trees and the Propagation of Timber in His Majesty's Dominions*, is more typical of the Society's outputs.)

In 2010, the Society celebrated its 350th anniversary, commissioning various commentators to pick out bits of its rich history. In his chapter, Simon Schaffer (2010) referred to an episode in 1772 as the Society was becoming more assertive in its engagements with politics. A spate of lightning strikes had prompted the Board of Ordnance, who had responsibility for storing gunpowder, to ask the Society's advice about lightning conductors. The committee established to respond included both Henry Cavendish and Benjamin Franklin. Cavendish became famous for experiments on gases and the density of the Earth. Franklin had already achieved fame as an experimenter on lightning and would go on to be a central figure in US politics. He adopted the 'age of experiments' brand for his own era and was referred to as the modern Prometheus (Schaffer 2010, p. 142) (a phrase that would come to provide the subtitle for Mary Shelley's *Frankenstein*) for his work in understanding and appearing to control lightning.

Institutions demanded certainty from the Royal Society, and Franklin was willing to provide it, even though other Fellows disagreed. Benjamin Wilson, the leading dissenter in this case, assembled and publicly performed his own experiments to demonstrate the advantages of flat-topped lightning rods over Franklin's spiked ones. The Royal Society's internal ructions were reported in the press, and its particular model of public trust began to wobble. The experiments that were supposed to speak to matters of fact without intermediation seemed to be causing greater uncertainty. Schaffer (2010) describes how the variety of experiments that took place, with varying degrees of control because of the reliance on a natural phenomenon, were contested to such a degree that the conclusions boiled down to a question of credibility – who can be trusted? (Schaffer 2010).

Scientists express constant frustration with a lack of action on climate change. They would all agree with David Hume that one can't turn an 'is' into an 'ought', but the debate often confuses the science with the policy. In a later chapter of the same retrospective, climate scientist Stephen Schneider reflects on the challenge of credibility that still remains: 'What constitutes enough credibility to act is not science per se, but a subjective value judgement on how to gauge risks and weigh costs' (Schneider 2010, p. 427).

The Royal Society finds itself stuck between the 'is' and the 'ought'. It is unsure if its role is merely to speak to the science or to use the science to speak out on public affairs. This lack of clarity is not a bad thing; it allows the Society to respond in multiple ways to a range of issues. The Fellows of the Society, elected by other Fellows on the basis of scientific accomplishments, express a range of views on the Society's role, as they do on most matters. The Society plays a role in organising such views and arranging their credibility. Although its motto captures the assumption that science speaks for itself, the Society's engagements with policy are, in practice, increasingly sophisticated. Many of its Fellows and staff are sufficiently progressive to keep ahead of trends in science policy. The Society takes credit, for example, for initiating the Public Understanding of Science movement in the 1980s and then the move towards Upstream Engagement in the 2000s (Gregory and Lock 2008). The UK government still comes to the Society in search of what Schaffer (2010, p. 144) calls 'unequivocal decision', but in most cases the institution is able to reframe their requests rather than simply responding. Nevertheless, the Society is unashamedly a scientific institution, and it displays a formal scientism, in Brian Wynne's sense of that term, presuming that 'science has natural sovereignty over public meanings' (Wynne 2014). When it comes to issues such as geoengineering, therefore, the Society has to actively resist its instincts to close down the relevant issues.

Experts and public policy

The growth of political awareness about climate change has exacerbated a tension between expertise and democracy, but this tension is not new. John Dewey pointed in the 1920s to the problems of governing technologically advanced societies relying on specialised expertise. As Frank Fischer describes, the rise of expertise parallels the rise of an assertive public in a strange paradox. While Dewey imagined an ideal scenario in which experts would inform an increasingly vocal democracy, we would seem to have ended up in a situation in which experts are cut off from democracy (Fischer 2000). Fischer quotes Harvey Brooks, US physicist and science adviser, who argued that 'much of the history of social progress in the twentieth century can be described in terms of transfer of wider and wider areas of public policy from politics to expertise' (Brooks 1965, p. 72).

Brooks does not see this as a result of power-hungry scientists, but rather the 'instinct of statesmanship . . . to turn intransigent problems over to "experts" or to "study groups"' (Brooks 1965, p. 72). Climate change is a perfect example of this dynamic. Because it is one of the dominant science-policy issues of our time, organisations such as the Royal Society fixate on climate change but find themselves occasionally paralysed in coming to terms with it. The definition of the problem is so unavoidably dependent on science, and yet the solutions on offer demand joined-up political action, leaving science relatively impotent. Discussions of climate change quickly descend into differences of opinion about the appropriate role of a national academy of sciences. Social scientists have demonstrated that advisory scientists are not doing '"science" in any ordinary sense, but a hybrid activity that combines elements of scientific evidence with large doses of social and political judgement' (Jasanoff 1990, p. 229). But many people at all levels of the Royal Society would still like to believe that their authority comes from a strict separation of science and politics. Many Fellows of the Society would agree with the popular version of the Society's story provided by science writer John Gribbin:

> The Royal Society itself, although allegedly founded on Baconian principles, certainly never took upon itself any role in the practical application of science to the immediate direct benefit of humankind; if anything, it did the reverse, encouraging speculative investigation of the world by people interested in knowledge for its own sake, not for its practical utility.
>
> (Gribbin 2006, p. 75)

The caricature here resembles the Grand Academy of Lagado described by Jonathan Swift in *Gulliver's Travels*, a satirical institution whose esteemed Fellows were obsessed with pure science to the exclusion of all practical knowledge (including 'extracting sunbeams out of cucumbers'). In his valedictory address before handing over the presidency to Martin Rees, Robert May took issue with Gribbin's description. Lord May had previously been the government chief scientific adviser, and under his leadership the Society became unafraid of entering political battles. At times, their engagements were ill-advised. Richard Horton, editor of *The Lancet*, was scathing about what he regarded as the Society's underhand and illegitimate meddling in a scientific debate about the health risks of genetically modified foods (Horton 2003). Horton accused the Society of targeting a scientist, Arpad Pusztai, who had published a paper in *The Lancet* that hinted at a health risk from genetically modified potatoes. In a later editorial, Horton advised Rees, then president-elect, to ask himself, 'What is the Royal Society for?' (Horton 2005).

Rees is a cosmologist by background. His tenure showed no sign of a relaxation of the Society's policy activities, although the mode of engagement, supported by a growing science policy staff, had been less clumsy than that of his predecessor. The Royal Society is an institution that offers myriad answers to Horton's question, depending on who asks, who answers and in what circumstances. The Royal Society can variously be seen operating as an advisory body, a research funder, a public relations outfit, a journal publisher, a conference facility and a gentlemen's club (with a growing but still small female membership).

Time spent inside the Royal Society reveals the gulf between the desired public image of the institution, in which, following Charles Lindblom's terminology, a synoptic view of issues is required before decisions are made and the reality, in which the various parties involved are 'muddling through' (Lindblom 1959; also see Stilgoe 2012). The Society presents an image of scientific rationality, with its Fellows as the public face of its reports, but much of the work and policy expertise come from the staff operating backstage.[3]

'Not completely stupid'

Following the publication of Crutzen's (2006) paper, discussions about geoengineering among scientists, particularly in the USA, grew. At the same time, political and policy bodies began to raise the issue. Some, including US right-wing think tanks such as the Heritage Foundation, the American Enterprise Institute and the Heartland Institute, did so enthusiastically. The Convention on Biological Diversity (UNEP 1992), meanwhile, was the result of a need to regulate experiments that were taking place on ocean iron fertilisation (OIF). These experiments had been conducted for decades, motivated by a desire to find out more about complex ecosystems (e.g. Coale *et al.* 1996) (see also Chapter 5). But since 2002 a handful of companies had begun to claim that OIF was a sufficiently reliable method of carbon capture to justify selling carbon credits. Tests had begun in the Pacific with this in mind. Other start-ups had begun to construct 'green' investment schemes based on claims about biochar as a means of carbon sequestration, arousing the suspicion of carbon dioxide removal (CDR) researchers.

The National Academies had spoken out on the issue more than a decade before as part of their report on the greenhouse effect (NAS 1992). The National Academies, encompassing the National Academy of Sciences, the National Academy of Engineering, the Institute of Medicine and the policy arm, the National Research Council, are the US equivalent to the Royal Society and its sister academies, but their role is subtly different. Set up by Abraham Lincoln during the Civil War, they are more intimately connected with government. Their role is to respond to the demands of the policymakers who are their paymasters, which gives them larger budgets but less independence. The Royal Society receives relatively little government funding for policy work, which means it can set agendas rather than responding to those of others. The National Academies' chapter on geoengineering looks at reforestation, OIF, and what they call 'screening out some sunlight' using dust in space or the stratosphere (NAS 1992, p. 447). In hindsight, their report displays a tin ear for the politics of climate change and geoengineering. They downplay the uncertainties and use chaos as a reason *to*, rather than not to, 'tinker' (NAS 1992, p. 435). The fast-moving debate on climate change meant that while the US report served an important purpose at the time, its value was rapidly overtaken by other studies.

The question of geoengineering had drifted in and out of the Royal Society's attention over the early years of the twenty-first century, raised sporadically by Fellows with interests in climate science. Geoengineering had, in the form of afforestation, been hinted at in a 2001 report on land carbon sinks (Royal Society 2001). In 2004, the Society had declined to be involved in a join US–UK workshop on geoengineering, a collaboration between the Tyndall Centre for Climate Change Research and the Cambridge–MIT Institute.[4] This meeting was, for some UK scientists, the first airing in a respectable, open scientific forum of some ideas that had, according to one attendee I interviewed, seemed as though they '*were going to be bonkers*'. The meeting exposed UK scientists to new people as well as new ideas. Lowell Wood gave a talk at that meeting about solar radiation management (SRM) in space and the stratosphere which, in the words of one scientist, '*scared the shit out of most of the people present . . . partly the boldness, the willingness to conceive of doing big things to the planet and partly also because we knew where he was coming from*'. For the Royal Society, geoengineering maintained an association with '*nightmarish 1960s technocratic scientists*' in the USA, according to one person close to the study.

Arguments for the Society's speaking on the issue were at first closed down. Other Fellows felt that even if the Society were to criticise the current crop of geoengineering proposals, to speak about them would be to lend them more legitimacy than they were worth. The Society had to judge whether its activities would take the heat out of the debate or further fan the flames. Meanwhile, it became clear that geoengineering was receiving plenty of oxygen on the margins of policy and politics.

Organisations with interests in stretching or breaking the political consensus on climate mitigation had begun to discuss geoengineering as an alternative or complementary approach. In the USA, right-wing think tanks such as the Heartland Institute and the American Enterprise Institute were talking up the prospects for geoengineering, with the latter announcing in June 2008 that 'a growing number of climate scientists believe that there may be only one possible answer . . . a concept known as "geoengineering" . . . most scientists who have studied the idea believe it is likely to be feasible and cost-effective' (AEI 2008).

Newt Gingrich, a leading Republican and former Speaker of the House, took things further that same month: 'Geoengineering holds forth the promise of addressing global warming concerns for just a few billion dollars a year. Instead of penalizing ordinary Americans, we would have an option to address global warming by rewarding scientific innovation' (Gingrich 2008).

The Council on Foreign Relations, a non-partisan foreign-policy think tank, held a workshop in May 2008 starting with a conventional (but, as described in the previous chapter, highly speculative) scenario of 'unilateral geoengineering' (Ricke *et al.* 2008).

At the same time, rumours had begun to circulate that the Russian Academy of Sciences was keen to refer to geoengineering in the statement of the G8+5 science academies, representing the views on climate change of the leading scientists from the world's richest nations. Yuri Izrael, a Russian scientist involved with the negotiations and with connections to both the Intergovernmental Panel on Climate Change (IPCC) and Vladimir Putin, had proposed some preliminary SRM experiments and was forthright in his enthusiasm for geoengineering (Sinitsyna 2008). In the end, this statement included an equivocal comment that 'there is also an opportunity to promote research on approaches which may contribute towards maintaining a stable climate (including so-called geoengineering technologies and reforestation), which would complement our greenhouse gas reduction strategies' (Royal Society 2008). With the looming December 2009 Copenhagen Climate Change Conference, at which participants would attempt to build a binding global agreement on climate change upon a fragile political consensus, scientists and policymakers were keen to prevent the emerging geoengineering discourse from destabilising the discussions.

The Department for Environment, Food and Rural Affairs (Defra) were keeping an eye on these developments on behalf of the UK government, viewing this initially as an issue to do with pollution in the environment rather than climate change, which was the responsibility of the Department of Energy and Climate Change. Defra's chief scientific adviser, Bob Watson, a previous chair of the IPCC, suggested to the Royal Society that it was an issue worthy of their consideration.

Once a person or institution decides to pay attention to an issue, activities that are under it are suddenly more apparent. In the summer of 2008, geoengineering appeared to be '*coming at us from a number of angles*', according to one Royal Society employee. Geoengineering was also circulating within the Society's journal *Philosophical Transactions*. A special issue of the world's oldest scientific journal, focussing on geoscale engineering to avert dangerous climate change, was due out in October 2008. Some of the papers had begun life at the scientific meeting in 2004. Two papers argued in favour of further OIF experiments. Others discussed cloud brightening or stratospheric aerosols. Brian Launder, one of the issue's editors, would go on to suggest to the House of Commons Innovation, Universities, Science and Skills Committee that the government should fund field trials of such geoengineering approaches (Fleming 2010, p. 261). One working group member had begun to detect '*a degree of rehabilitation*' in geoengineering research. On closer inspection of some proposals that had seemed outlandish, this scientist saw that '*many of these ideas were not completely stupid, and worthy of investigating*'.

In this way, various individuals and institutions were able to justify a growing interest in geoengineering with reference to the activities and research of others.

Even if scientists would rather the technology were never developed, they justify research by the need to understand the intricacies and implications of technological development and deployment by others. The presence of this research then may make further research, policy consideration and even technological development more likely. This spiral of justification takes on some of the features of the arguments put forward in support of biological weapons research during and after World War II. In this case, countries agreed that 'defensive research', in response to the perceived or actual activities of others, was more legitimate than 'offensive research', but it was impossible to completely delineate one from the other (Dando *et al*. 2006).

In July 2008, the Council of the Royal Society discussed and approved a paper proposing a study on 'sustainable geoengineering'. The justification ran along lines that had by this stage become familiar: policies to tackle climate change seem to be ineffective; some form of geoengineering somewhere seems not just possible but likely, given its relative cheapness and the scale of political desperation; there is therefore a need to inform the debate and advise on regulation. The Council took the decision to assemble a full working group for a year-long study, the Society's highest possible level of engagement with an emerging issue.

The Council decided to proceed relatively quickly, in time to report before the Copenhagen Conference in December 2009. But this didn't stop the sense of ambivalence felt by many inside and outside the institution. There was a general sense that the Royal Society report would hasten the lifting of whatever taboo remained after Crutzen's (2006) intervention. Opinions differed over whether this was a bad thing. One working group member detected '*varying degrees of support for the idea that the climate negotiations were clearly failing, and therefore some kind of alternative or supplement to the approach that had thus far been taken to those negotiations was required*'.

Scientists with long commitments to the science of climate and the implications for mitigation to which it seemed to clearly point had previously expressed concerns about the growing attention given to climate *adaptation*. What seemed to be a taboo on adaptation had started to lift (see Pielke *et al*. 2007), and some of the Society's Fellows were, according to one Society employee, worried that

> '*to throw in a further spoiler to the mitigation story and to give it any air-time would be difficult . . . An early impression from a lot of the group members was that if you start talking about this and get too excited by it, then you end up giving up on everything else*.'

This concern about adding the Society's authority to existing speculation was met with arguments that the debate was happening anyway. Scientists who would once have treated geoengineering proposals with disdain argued that it was better for the debate to be informed and open in the pages of scientific journals and reports than hidden in the corridors of power (see, e.g., Lawrence 2006). This transition was reflected in the Society's working group. One employee describes an '*early, nervous working group meeting*':

> *'As it evolved, it became clear that it [the issue of geoengineering] needed to be out there . . . There was a split in the group. There were those who "got it" and were vociferous in their opinion of the need to have this contained, and all of these fanciful ideas being put into some kind of structure . . . And then there were those who were more reluctant . . . about whether we should consider it at all.'*

The first set of names proposed for the working group comprised entirely scientists and engineers who had previously been involved with geoengineering. However, following the appointment of a chair, John Shepherd, and a small core group of scientists, including Ken Caldeira, the final working group was chosen to reflect the broad range of disciplines that had already considered geoengineering. Royal Society staff were key to stretching the definition of expertise that would inform the choice of people. Researchers such as Caldeira and Keith, who had already published research on geoengineering, were joined by self-confessed '*outsiders*', scientists with a broad interest in climate change, a lawyer, Catherine Redgwell, and the social scientist Steve Rayner. The group decided they needed expertise from economics, so Gordon MacKerron, director of the longstanding Science Policy Research Unit at the University of Sussex, was invited to join. The group discussed whether to invite an ethicist on board, but decided instead to run an 'ethics panel' meeting with three external experts. Brian Launder, who had been an editor of the special issue of *Philosophical Transactions*, was the only fellow of the RAEng on the group. When discussions began about the Society's study into geoengineering, a proposed partnership with the RAEng was quickly dismissed on practical grounds. It was felt that it would take too long to get agreement between the two academies on the selection of people, division of workloads, budgets and text. But the Royal Society's framing of its study and its subsequent defences in the face of behind-the-scenes lobbying from engineers also reveal a nervousness about the engineering part of geoengineering.

In March 2009, a report from the House of Commons Innovation, Universities, Science and Skills Committee considered geoengineering as part of an investigation of 'turning ideas into reality' in engineering. The committee is one of the House Select committees, which play an important role of parliamentary scrutiny of government, but they are typically under-resourced. They rely on a judicial model in which members of Parliament question 'witnesses', so the impressions that they provide of issues can be patchy. Nevertheless, the committee discovered an important disconnect between government departments on geoengineering. The committee, searching for a policy position on geoengineering, heard enthusiasm from the excitable science minister, Lord Drayson. But the official line from the minister of state for energy and climate change, Joan Ruddock, was that geoengineering proposals should not allow policymakers to be distracted from the task of mitigating climate change by reducing emissions. The committee, failing to grasp the uncertainties of geoengineering proposals and enthusiastic about seizing the 'opportunity to restructure the economy by building on the existing substantial strengths of UK engineering' (House of Commons 2009, p. 6), demanded a more pro-active policy approach.

Framing the report

The performance of scientific advice places great emphasis on the processes – working group meetings, consultations, submitted papers, peer review and such – that are supposed to ensure that institutions 'get the science right' (Lentsch and Weingart 2011). But studies of expertise on contentious issues suggest that the answers that come out crucially depend on the questions that get asked (Stirling 2008). The way in which expert advice is sought and framed matters as much as the way in which it is assembled and presented.

Early discussions among the working group built a constituency around the importance of the project and its approach, within which the working group could have open arguments: '*We had some very heated arguments, primarily on technical matters. But there was a very broad consensus from the outset about what we were trying to do and how to do it.*'

In the constitution of the working group, the open-mindedness of the chair, the approach of the staff and the activities undertaken, the Royal Society's (2009b) geoengineering study represented a substantial opening up of the institution's normal advisory process. As with British science policy more generally, the tendency had been for the Society to operate in a highly linear, technocratic mode. Science would speak truth to power, it was assumed. This model began to unravel in the 1990s with controversies of bovine spongiform encephalopathy (mad cow disease) and genetically modified crops, leading to evolution and experimentation taking various forms (Millstone and van Zwanenberg 2001; Stilgoe *et al.* 2006). The Royal Society's 2004 report on Nanoscience and Nanotechnology, undertaken in partnership with the RAEng (RS–RAEng 2004), had set the tone for subsequent working groups in its openness to considerations and actors that went well beyond the scientific. Political awareness of upstream concerns led to this report having the subtitle 'opportunities and uncertainties', as opposed to the more technocratic, more certain alternative: 'benefits and risks'.

However, to turn the complex sociotechnical issue of geoengineering into a doable science policy study, the frame of reference had to be narrowed. This allowed one working group member to present the report as an example of what he called '*classical science policy*':

> '*All opinions on this are valid . . . as far as possible establish the scientific facts as a basis for a wider discussion . . . and then let battle commence, but trying not to let opinion influence the things that you could say something about in a factual way.*'

For all that inputs to the scientific advice of the Royal Society were opened up, its cultural and institutional limitations constrain its ability to provide what Andy Stirling (2010) calls 'plural and conditional' outputs. The Society's instinct is to speak with a unitary voice, the voice of science, pretending towards what Roger Pielke (2007) has labelled the 'pure scientist' mode of advice. The ambition to adopt a more relevant, more engaged 'honest broker' role, which was shared by some of the more progressive Fellows and many staff, was often met with intransigence by those Fellows with less interest in or experience of political processes.

The public engagement process incorporated in the study, which I helped to design, was a late addition. It was commissioned long after the study had begun and its report had been framed; the idea was to use a small budget of a few thousand pounds to run a series of focus groups and conduct a short survey. Various rationales were put forward for this work, with some keen to use it as a way to explore the relationship between geoengineering technologies, public opinion and public action on climate change. The engagement exercise attempted to test the question of whether geoengineering presented a 'moral hazard'.

The idea of a moral hazard had, by the time of the Society's report, become central to governance discussions of geoengineering. A moral hazard, as understood by economists, is a situation when one party is insured against their own risk-taking and so becomes more likely to take risks. The reasoning is that if geoengineering is an insurance against the effects of climate change, the world will become less likely to mitigate. If the hazards posed by geoengineering are moral as well as environmental, these may start to be felt much earlier than the direct risks that might come from the testing or use of a technology. Indeed, they may already be happening, if the mere suggestion of geoengineering causes policymakers to relax.

The moral hazard argument was initially used by critics of geoengineering as an argument against endorsing, developing or even researching geoengineering proposals. But it has quickly become an empirical, behavioural question for geoengineering researchers with multidisciplinary ambitions. They want to know whether citizens and consumers would in fact change their behaviour in the face of new options for climate change, or whether current approaches to mitigation have been shown to be so ineffective that nothing could make them worse. Given the complexity of climate policy, the moral hazard is a dramatic over-simplification in which climate policy is presented as an either/or and mitigation rests on a fragile political consensus. The previous experience with climate adaptation, discussed above, is instructive here.

A number of philosophers have taken apart aspects of the moral hazard debate. As Ben Hale (2012) has pointed out, it is not clear where the moral hazard is expected to fall. Are we talking about politicians losing interest in mitigation, consumers making unsustainable choices, researchers turning attention away from climate science, or engineers giving up on clean technology? We can predict that at some point new technological options would lead to new behaviours and a redrawing of political fault lines, but the moral hazard argument as currently imagined makes some assumptions about geoengineering that are unwarranted. In the language of economics, we might ask about the credibility of the insurance that geoengineering is said to offer. A technological determinist would see geoengineering as inevitable, if not available in the near future. To even raise the possibility of geoengineering is therefore to draw up an insurance policy that threatens to radically change behaviours in the face of climate change. However, as is my argument in this book, the uncertainties are profound. Scientists, by claiming to know about geoengineering, may, if they are not careful in their framing, be contributing to the pretence of insurance.

The problem is not, as is often presumed, the discussion of geoengineering per se, but rather the presentation of geoengineering as if it is likely. To the extent that scientific authority itself gives emerging technologies legitimacy and is typically assumed to close down rather than open up options (Stirling 2008), scientists should indeed ask themselves 'to speak or not to speak' (Lawrence 2006). But they should also consider carefully how to frame their speech. The Royal Society working group managed to maintain a remarkable reflexivity about its own statements on geoengineering, but they were not helped by a dominant idea of a moral hazard that itself makes geoengineering appear concrete. As I described in Chapter 2, the sociology of expectations reveals that promises made about emerging technologies contribute to the shaping of particular futures. The moral hazard argument can be seen as part of a set of promises about the power of geoengineering. Following Mike Fortun (2005), we can therefore suggest the need for an 'ethics of promising' in the claims made about geoengineering. Those, including myself, who support and practise deliberative public engagement around emerging technologies must also be aware that theses processes can themselves reinforce particular visions of the future, even while attempting to open them up.

The focus groups conducted by the Royal Society revealed an instinctive distaste for uncertain, high-leverage proposals, such as stratospheric particle injection, a finding that would be deepened by further qualitative social research (e.g. Macnaghten and Szerszynski 2013). The nuances of this concern did not make it into the Society's final report (Royal Society 2009b). The short summary of the process mentioned instead the counterintuitive finding that people seemed more driven to mitigation action by geoengineering, rather than becoming more complacent as the moral hazard argument would suggest. The suggestion was that people would be scared into action by the realisation that scientists were considering such desperate alternatives. The public engagement work came too late to influence the direction of the report in any substantial way, except that there was some discussion that the findings would be useful for correcting public perceptions of risk from geoengineering technologies that appear to have been detected.

Despite early efforts from some on the working group to thread social, ethical and political questions through the whole report, the discussion on governance was eventually placed in its own chapter (Royal Society 2009b). The understandable explanation is that time constraints prevented a full consideration of the governance intricacies of the different options proposed or of the questions about responsibility in research that are the focus of this book. But the project of true multidisciplinary technology appraisal is also complicated by cultural assumptions about the relationship between science and governance that are played out in the operations of the Royal Society, just as they are in the production of the geoengineering report. A majority of the working group thought that 'technical' and 'social' considerations should as far as possible be kept separate. This is not to downplay the novelty of the approach taken. In the constitution of the group and the production of the report, the Royal Society stepped well outside its scientific comfort zone and, in doing so, placed scientific evidence and expertise alongside rather than above other considerations. One working group member described how '*everyone was surprised in the end by the extent to which the report stepped beyond the Royal Society's usual remit. I think even the authors were surprised with "governance", being in the title of it, or the subtitle at least*'.

However, the separation of science from governance allowed for some momentary relapses in the report. One group member claimed that '*the worst thing that came out of that [separation] was the blob diagram*'.

The 'blob diagram'

An important part of staging a performance of scientific advice, to follow the theatrical metaphor, is calculating which parts of a complicated, heavyweight report should be projected most forcefully to the back of the auditorium. Reports are presented and press releases are written with some idea of what, from the mass of information and conjecture, the audience is expected to take home. For the Royal Society's (2009b) report on geoengineering, one picture, referred to by those close to the study as the 'blob diagram', would come to represent their overall assessment of geoengineering (Figure 4.1).

The blob diagram contains no more information than found in the table of numbers on the report's previous page. And yet, as historians and philosophers of science have discussed, even if diagrams can in principle be substituted for words, they can play very different roles in representation, argument and communication.[5] In their abstractions, they are, as Deleuze and Guttari (1988) recognise, another stage of detachment from the material world. Diagrams, like models, force certain choices about what to exclude, what to include and how. The choices behind such diagrams relate to the purposes they are trying to serve, but in the case of the blob diagram, these purposes were multiple and conflicting.

The diagram was a late addition to the report, decided upon by the working group during a hot London summer afternoon. Most within the group felt that a visual representation of the data would help inform a policy debate on geoengineering options, although there was disagreement about what should go where. The challenge was to represent four dimensions of criteria on one diagram – affordability, effectiveness, safety and timeliness (how quick or slow each method would be to deploy). The decision was taken to place affordability on the *x*-axis and effectiveness on the *y*-axis and to represent safety by colour and timeliness by the size of blob (although, as one working group member later admitted, safety could equally have been plotted against effectiveness).

The movement of some of the options between drafts of the diagram suggest that their final locations were strongly dependent on framing assumptions. The idea of 'surface albedo', which was initially plotted as both affordable and safe, was switched to become unaffordable and unsafe following a discussion of the scale at which such an intervention might be expected to make a difference.

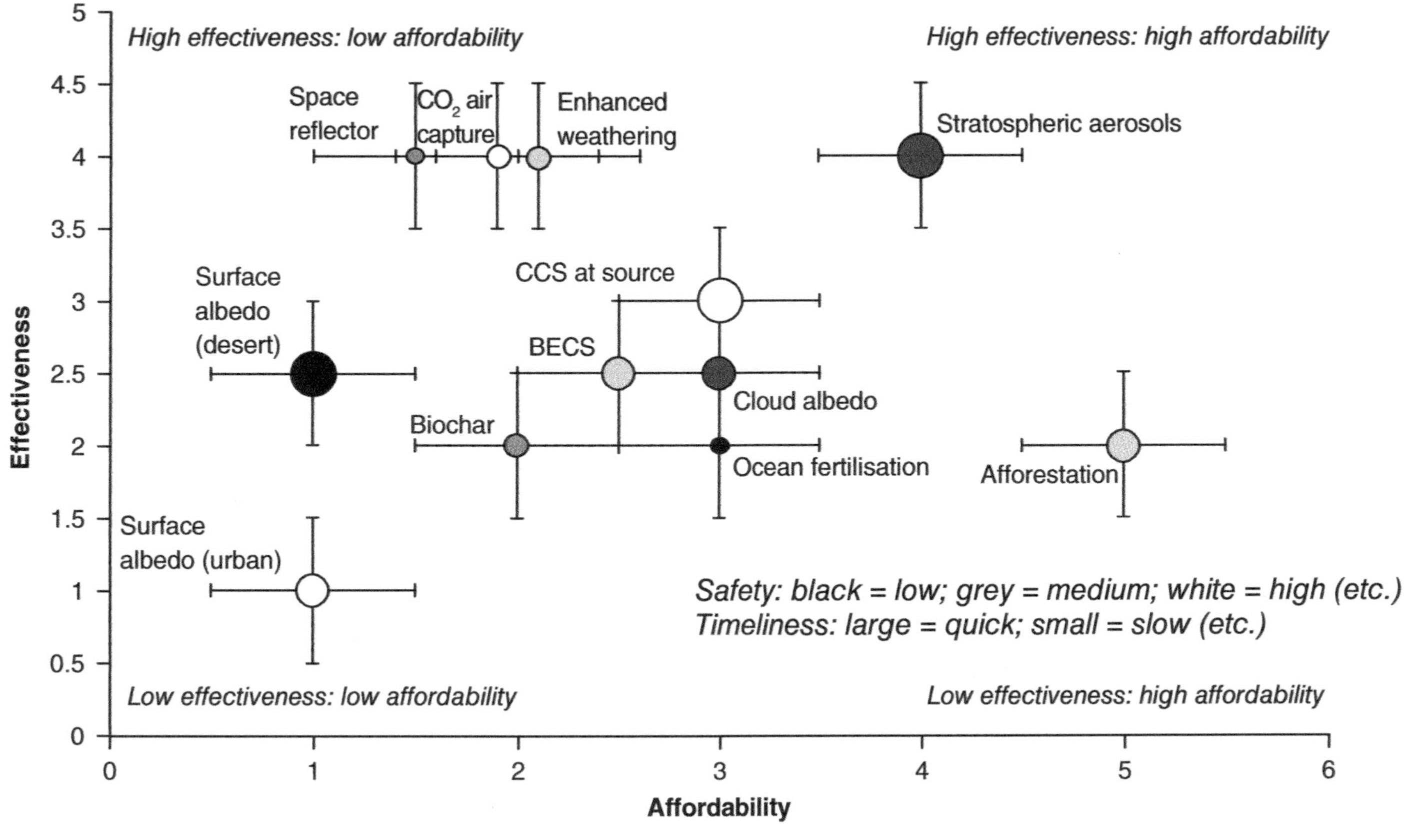

Figure 4.1 Preliminary overall evaluation of the geoengineering techniques.

Source: Reprinted from Royal Society (2009b, Figure 5.1). Note: BECS, bioenergy with carbon storage; CCS, carbon capture and storage.

Geoff Brumfiel, a reporter for *Nature*, called the diagram a 'bizarre phase space' (Brumfiel 2009). Others were quick to point out oddities such as the single decimal place on the units of only the *y*-axis. Error bars were used to represent uncertainty, but they were all the same size, and the dimensions of safety and timeliness had no error bars at all. The report explained, perplexingly, that 'the error bars are not really as large as they should be, just to avoid confusing the diagram' (Royal Society 2009b, p. 54). Even though the diagram managed to include four dimensions of assessment, some members of the group pushed for a rethink. One told me that the diagram was '*overly narrow in the criteria that it used, but also conveyed a greater sense of certainty about the information it contained than was warranted*'.

One could, as Tim Kruger has done, take apart each dimension. The affordability axis, for example, is based on cost per unit of reduction of radiative forcing. The effectiveness axis similarly presupposes that the important criterion for success is the reduction of global average temperature. Kruger asks, 'Why temperature and not precipitation, or human suffering, or food supply, or protection of biodiversity, or ocean acidification . . . or a host of other potential metrics?' (Kruger, forthcoming).

To illustrate the effects of these choices, Andrew Maynard (2009), an environmental risk scientist and blogger, replotted the diagram, putting safety and effectiveness on the two axes (Figure 4.2). The data remain identical, but stratospheric aerosol is no longer at the top right.

Statements, diagrams and conclusions from institutions such as the Royal Society are sometimes, if politically convenient, offered or received as though they are the final word on a subject. In reality, such assessments are always unfinished conversations. Some working group members involved with the Royal Society report admit the subjectivity of the blob diagram and other assessments within the report. They recognise, for example, that safety is itself subjective, highly contingent and multidimensional (I was told that '*science does not have an SI unit for safety*'). One defence offered was that the diagram was merely a '*tool for kickstarting conversations about SRM*'. But a single purpose behind the representation was never resolved. For some in the working group it was a way of indicating which technologies we should be most concerned about in governance terms. For others it was about deciding priorities for natural science research. One reviewer of the report went even further, expressing the need to decide priorities for the development of workable technologies.

A commentary by philosopher Stephen Gardiner (2011) explores some of the explicit and implied ethical considerations in the report. Gardiner is right to suggest that the framing of the report suggests certain ethical assumptions and lacunae. He notes, for example, that the possibility of moral hazard and technological lock-in are treated as empirical questions, claiming that there is little evidence in support of either. It is certainly interesting that greater certainty is demanded on social and ethical questions than on technological ones.

But there is a danger of reading too much rationality into its arguments. Such reports are trapped in a model that demands a consensus statement but deals with issues and in timescales that make consensus impossible. Beneath the surface,

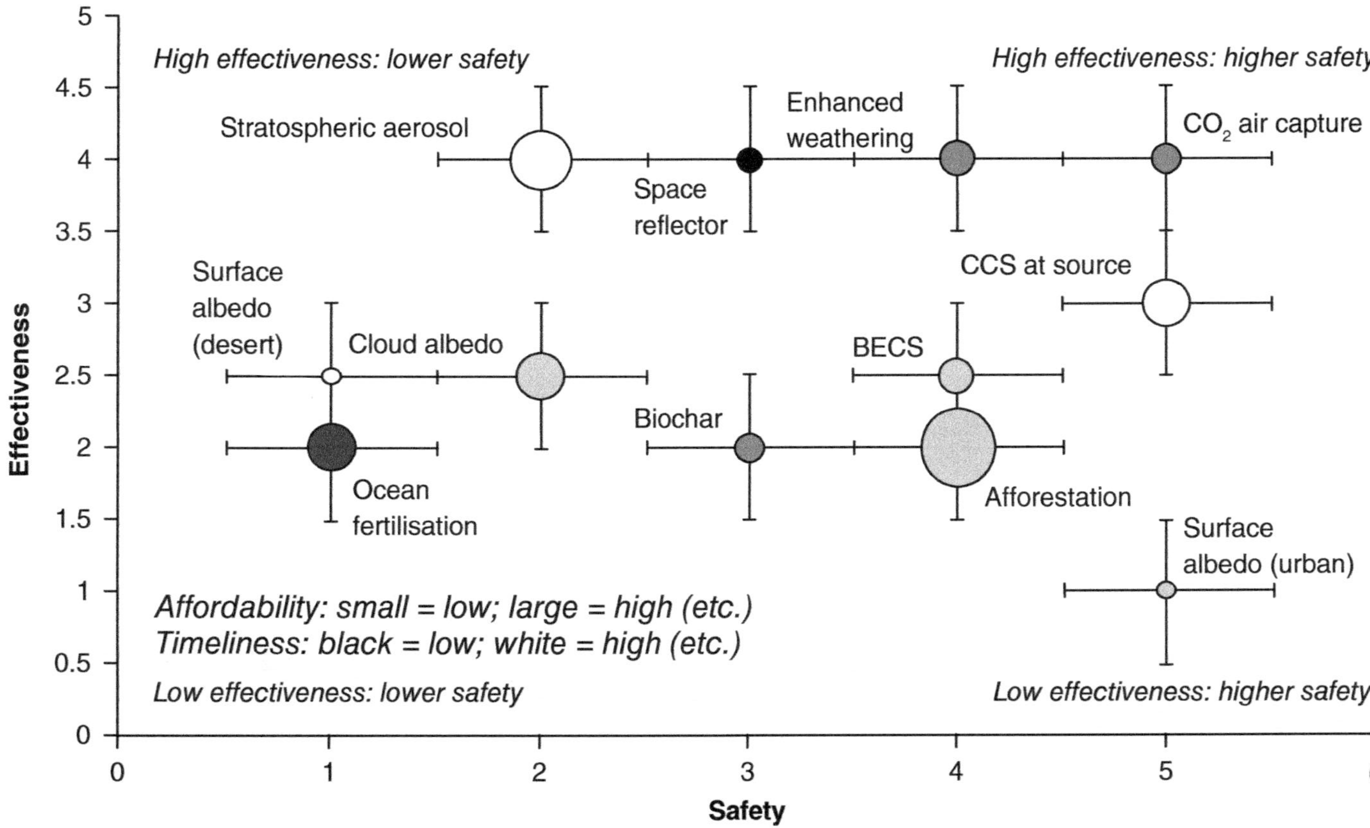

Figure 4.2 Displaying estimated effectiveness versus safety for twelve geoengineering approaches.

Source: Based on data in the Royal Society (2009b) *Geoengineering the Climate* report, replotted by Andrew Maynard (2009) and reprinted with permission. Note: BECS, bioenergy with carbon storage; CCS, carbon capture and storage.

such reports are always riven with little contradictions. For example, on matters of economics the report argues that 'quite apart from the limited capacity of simple economically focussed cost–benefit climate impacts assessment models to provide policy-relevant results, analyses of whether to do either geoengineering or emissions mitigation are inappropriate' (Royal Society 2009b, p. 44). The report also critiques economic 'appraisal optimism' (Royal Society 2009b, p. 44). Similarly, the report foregrounds the idea that 'technology to do [geoengineering] is barely formed, and there are major uncertainties regarding its effectiveness, costs, and environmental impacts' (Royal Society 2009b, p. ix). But this reasoning doesn't reach across to assessments such as the blob diagram.

At times in the report the language slips, and conflicting views show through the cracks. Stratospheric aerosols, which emerge as the cheapest, most effective and quickest option for geoengineering, are described at one point as 'the most promising' of the SRM proposals (Royal Society 2009b, p. xi). Elsewhere, this option is described as the 'nearest approximation' to 'an ideal method' (Royal Society 2009b, p. 49) currently in play. According to Bellamy *et al.* (2012), the calculation of stratospheric aerosols as a best option, even while disagreement was rife within and around the group about what 'best' meant, contributed to a closure on this in subsequent geoengineering research and policy discussions.

One of the Royal Society report's external reviewers suggested that the uncertainties surrounding the unintended consequences of stratospheric aerosols justified ruling it out as an option entirely, rather than making it a research priority. This reviewer saw the blob diagram as enormously problematic, recommending that it be cut on the grounds that a full consideration of uncertainties made quantification impossible. As Ted Porter (1996) has described, the whole idea of cost–benefit analysis, of which the blob diagram is an example, presupposes that incommensurable criteria can be somehow made comparable through numbers.

Social science research conducted after the Royal Society's report reveals that when geoengineering options and other alternatives are assessed on a wider range of criteria, the uncertainties that surround them are revealed to be vast and multidimensional (Bellamy *et al.* 2013). Overlaid onto the original diagram (Royal Society 2009b, Figure 5.1), these might be represented as an error bar (or, more accurately, an 'ignorance bar') for stratospheric particle injection that stretches almost the full length of both axes, as well as in the original errorless dimensions of safety and timeliness.

Diagrams and numbers travel in ways that complicated ideas do not. The critical reviewer mentioned above warned the Royal Society team that their diagram would shed the nuances and contingencies that informed its creation. Nevertheless, the group proceeded, as one put it, '*with full recognition that it [the diagram] would take on a life of its own*'.

The blob diagram travelled quickly into the nascent geoengineering research community. In an early conversation, one of the Stratospheric Particle Injection for Climate Engineering (SPICE) team sketched for me his own version of it and explained their rationale for concentrating on stratospheric aerosols [pointing to the top right-hand corner of the sketch]:

> *'We're looking for things that are effective and cheap, looking for things in this region. You work your way up here and the first one you come to is particles in the stratosphere, and if you think of that as being effective then how do we get the particles in the stratosphere cheaply? Here you might do it with aircraft, here you might do it with missiles, here you do it with balloons, a tethered balloon. So, if you, as an engineer, are going to research something, what are you going to research first?'*

Despite the presence of the reductionist blob diagram, however, the Royal Society report was remarkably successful in managing to communicate the complexity of the issues under consideration.

The finale

The report was due to launch on the first of September. In the weeks before, however, the Royal Society was told about the imminent launch of a report from a smaller learned society in London, the Institute of Mechanical Engineers. Their intention to launch on the Thursday before the bank holiday weekend seemed calculated to steal the Royal Society's wind. While the engineers' report (IMechE 2009) pretends to be a form of technology assessment, it ends up as a thin advertisement for green engineering.

The engineers' three chosen technologies – artificial trees, reflective roofs and algae on the sides of buildings – present a benign view of 'soft' (Olson 2012) geoengineering and allow the report to include the sort of hi-tech graphics that appeal to news media. Their pictures of what one Royal Society employee called '*ping pong paddle carbon suckers*' lining motorways and nestled among wind turbines were readily picked up to illustrate subsequent geoengineering news pieces. But there was little serious reflection on the issues raised by more disruptive geoengineering proposals. The Society's concern was that the engineers' report threatened not only to satiate journalists' appetite for geoengineering stories, but also to gate-crash their own careful staging of a complex issue. As it was, the Society's report received plenty of attention.

If scientific advice is a performance, then a report's recommendations are its triumphant finale. There is rarely a twist in the tale. Recommendations are often an afterthought, and many scientists on advisory groups feel either unwilling or unable to offer precise messages to policymakers. In its 'truth to power' mode, the Royal Society takes up an aloof position, which makes it hard to engage with the specific needs and constraints of policy. Occasionally, the Society will speak back to the scientific community about its own responsibilities, but it is more common for working group reports to offer general suggestions for what others should do. The most clichéd of these is the ever-present call for 'more research'. We might find this unsurprising. As novelist Margaret Atwood (2003) puts it, asking scientists for practical recommendations is 'like asking ants what you should have in your back yard. Of course they would say "more ants".' This instinct is not one of greed, but is instead based on the assumption that answers are out there and uncertainties can be reduced. The working group shared a view that proper advice was impossible in the absence of solid knowledge, and knowledge was impossible in the absence of research.

The Working Group on Geoengineering knew from the start of their study that they would push for research funds. This was an area still regarded by some as untouchable, and researchers detected funders' nervousness towards anything that might have be seen as an endorsement. The Society requested a '10 year geoengineering research programme at the level of the order of £10M per annum' (Royal Society 2009b, p. xii). Working group members and staff subsequently told me that they were conscious of the need not to ask for too much. They did not want to divert substantial resources away from other areas, especially at a time of funding cuts across the UK.

The recommendation for funding was modulated with other statements about the need to govern this research, including establishing a *de minimis* threshold above which experiments would require additional scrutiny. The other recommendations each tell their own stories. The first recommendation is that mitigation should be stepped up, a message that would become the headline for the report's press release. This is followed by a rejection of the idea of using unproven CDR schemes like biochar and OIF to claim carbon credits (see Chapter 5). This recommendation, which sticks out from the others for its specificity, is a direct response to biochar lobbyists who were seeking formal recognition of their commercial potential in the run up to the Copenhagen Climate Change Conference.

The report also cemented the division between SRM and CDR, stating that they should in effect be evaluated separately. CDR was given cautious support, subject to research that demonstrated its efficacy, while SRM was regarded as more obviously potent, harder to govern and justifiable only in an emergency. Recommendations for the use of the technology were drawn up to encompass 'large scale experimentation or deployment' only (Royal Society 2009b, recommendation 4, p. 59). A recommendation to establish a 'code of practice for geoengineering research' (Royal Society 2009b, recommendation 7.1, p. 61) inspired the development of the Oxford Principles by a group including two of the working group's non-scientists and the subsequent creation of the Solar Radiation Management Governance Initiative, a forum for further discussion created by the Society in partnership with the Environmental Defense Fund and TWAS (The World Academy of Sciences, established for developing countries). (The Society's role in creating this initiative again marked a break from tradition. The norm was to launch reports and move on to new territory, rather than continuing to engage in emerging issues.) The Society also offered recommendations for international coordination of research, the urgent policy attention of the United Nations, and a process of public dialogue, which would become the Experiment Earth exercise. However, the report's major contribution was not its specific suggestions but rather the tone of voice with which it spoke.

The working group and staff of the Royal Society were conscious that their various audiences would read the report in myriad ways. The tone was apologetic: none of the working group wanted to be working on geoengineering; they thought it distasteful, dangerous or ineffective. In the report's first and last chapters and at its launch the group tried to convey that this was hurting them more than it was hurting us, the audience. Their proclaimed reluctance and caution in assessment were a clear signal that unthinking enthusiasm for a technological climate fix was no longer acceptable. The working group emphasised over and over again their commitment to mitigation, with the press release for the report taking a strident headline: 'Stop emitting CO_2 or geoengineering could be our only hope' (Royal Society 2009a). This form of scientific ultimatum is the same justification that Crutzen (2006) had used for speaking out a few years earlier.

Martin Rees, the Society's president, wrote a foreword to the report that added an additional metaphor – geoengineering as a 'plan B' for climate change. The 'plan B' phrase had begun to circulate in discussions of economic policy a few months earlier because of the uncertainties generated by the global financial crisis. As with other framings of geoengineering, the implications of this were never completely clarified. It was not clear, for example, whether a plan B would sit alongside the plan A of emissions reduction or if, following the use of the phrase in political debate, adopting plan B meant a radical change of direction, in effect dispensing with plan A entirely. Either way, this was not part of the story constructed by the working group itself.

In response to the report (or, more precisely, in response to a guess at what the report would contain, given the strict embargo), ETC Group issued a statement criticising the 'geoengineering enthusiasts' on the committee and recommending a 'ban on real-world experiments'. ETC speculated that a slippery slope would lead from growing enthusiasm or carelessness towards deployment: 'A yellow light can quickly turn green . . . Even the most careful computer models won't be able to predict what will happen if an experiment is scaled-up and moved out of doors' (ETC Group 2009a). In their accompanying document, rushed out to coincide with the Royal Society's launch, as well as taking issue with the assumed cheapness of geoengineering, they took a direct swipe at the Royal Society: 'the Royal Society's insular, opaque and languid process is incomprehensible at best, appalling at worst' (ETC Group 2009b).

This attention to the politics of science and technology is familiar from ETC Group's previous involvement in debates around transgenic crops, nanotechnology and synthetic biology. ETC (officially the Action Group on Erosion, Technology and Concentration) began life as the Rural Advancement Foundation International. They fuelled fierce opposition to genetically modified crops by accusing biotechnology companies of developing 'terminator seeds' that would produce a sterile next generation. They then intervened in the nanotechnology and synthetic biology debates to dramatic effect, releasing quick, well-researched reports that captured and shouted about the political issues behind emerging technologies. ETC are often scorned by scientists, even those who share some of their environmental values, for underhand tactics in their representation of science and its practitioners. Their model is to create big splashes with few resources, using acute political instincts. Following the argument that controversies provide opportunities for 'informal technology assessment' (Rip 1986), I would argue that the ability of ETC Group to bring attention and controversy to emerging technologies puts them among the most influential technology assessment organisations in the world.

The reception of the Royal Society (2009b) report elsewhere was substantially kinder. Most media reports picked up on the cautious tone. The *Financial Times* took the report as saying 'Hopes dashed for geo-engineering solutions' (Harvey 2009), while Reuters heard 'World must plan for climate emergency' (Reuters 2009). In most cases the message got out that research was needed, but for now geoengineering had been given a thumbs down as an alternative to mitigation.

Monsters and fairies

The Royal Society is a scientific institution to which many people would look for the authoritative, and singular, voice of science. The geoengineering report (Royal Society 2009b), by virtue of emanating from a scientific body, contributes to scientising an issue that is social and political in myriad ways. But as with other British science policy institutions, the Society has begun to rethink its political contribution – albeit constrained by conservative assumptions, inside and out, about its proper role. Building on a model of interdisciplinarity and caution that the nascent geoengineering community had already established, the Royal Society attempted to provide an open assessment of these proposals, drawing on a range of expertise.

The Society's concern was that, as the first national academy of science to devote a full study to geoengineering, it would give legitimacy to a highly speculative set of ideas that had the potential to radically destabilise political discussions about climate change. The report was taken to the Council of the Royal Society for approval, where it met opposition from those who had been, until that point, uninvolved in the discussions. The staff and Fellows of the Society were right to be concerned. For anyone looking at geoengineering, including critical social scientists, it is easy to quickly forget how strange this idea is. Alfred Nordmann (2006) talks about the 'uncanniness' of nanotechnology – a 'noumenal technology', undetectable by the senses. Although it is at the other end of a spectrum of scale, geoengineering technologies would be similarly uncanny. The idea of engineering the climate has no immediate comparisons with other engineering projects, and the effects and side effects would, as explored in subsequent chapters, in many cases be undetectable or undescribable by non-experts and impossible to completely represent scientifically. The danger is that when technologies are made thinkable, they become stabilised as normal, and expert assessments can lose sight of this strangeness. David Keith admits that 'it is a healthy sign that a common first response to geoengineering is revulsion' (Keith *et al.* 2010). The challenge is to keep hold of this while geoengineering research becomes normal. Keith has been honest about his own scientific enthusiasm, saying, 'We're hiding a genuine, and I think not-wrong joy in the fact that we understand something about the world that potentially gives us the ability to do these things' (quoted in NPR/TED Staff 2013). This enthusiasm is not about unalloyed benefits of a technology. As with synthetic biology, geoengineering has emerged as problem and solution simultaneously (Ginsberg *et al.* 2014).

The paper by Paul Crutzen (2006), the report of the Royal Society (2009b) and the subsequent growth of geoengineering research do not mean that the cat is out of the bag, the genie is out of the bottle or the horse has bolted (some of the many metaphors overheard in geoengineering governance discussions). The uncertainties surrounding these technologies, and the work required to make them real, are systematically downplayed. And while fully fledged technologies can be unruly, we underestimate our ability to put genies back in bottles, cats back in bags and horses back in stables. Geoengineering is less like a monster unleashed and more like the fairies in Peter Pan, which exist only as long as people believe in them.

The idea of a taboo presumes the existence of an unspeakable truth. Geoengineering is the technology that dares not speak its name. In reality, there is no technology that can do the things that geoengineering researchers hope or fear. Returning to Frankenstein, we should consider Bruno Latour's interpretation of the story:

> Just as we have forgotten that Frankenstein was the man, not the monster, we have also forgotten Frankenstein's real sin . . . Dr. Frankenstein's crime was not that he invented a creature through some combination of hubris and high technology, but rather that he abandoned the creature to itself.
>
> (Latour 2011)

For Latour, the moral of Frankenstein is to see technologies not as things that can be created as good or bad, but rather as works-in-progress. We therefore need to think about how scientists, scientific institutions and others might care for and nurture technologies. The emergence of geoengineering has opened up a space that is both scientific and political, which scientists find themselves navigating with greater or lesser confidence and ambivalence. In the next few chapters, using the SPICE project as a case study, I discuss how scientists have begun to consider their own responsibilities.

Notes

1 This description owes a debt of gratitude to James Wilsdon, who described an earlier report from the Royal Society, this time on nanotechnology, in similar terms (Wilsdon and Willis 2004), drawing on Hilgartner's (2000) analysis of expert advice as public drama. Wilsdon (a close colleague and collaborator of mine) was subsequently appointed to head up the Science Policy Centre at the Royal Society, an interesting case of theatre-critic-turned-director. He, along with colleagues Andy Parker, Rachel Garthwaite and Richard Heap, was largely responsible for encouraging the Society's open-minded approach to geoengineering and other issues.

2 There is a minor controversy about this. The Leopoldina – Germany's academy of sciences – was created in 1652, but it was not granted its official charter for another 30 years. The Royal Society was given its first Royal Charter by Charles II in 1662.

3 Because I am a former staff member of the Royal Society, this statement could be read as entirely self-serving. Having observed the Society's geoengineering work, however, I can

attest that it would have been substantially less nuanced, less relevant and less effective without the leadership of staff, in particular Andy Parker, who, after completing the study, went on to lead the launch of the Solar Radiation Management Governance Initiative.

4 For symposium report see Tyndall Centre for Climate Change Research (2004).

5 See the cluster of papers in *Biology & Philosophy*, 6(2), particularly Gilbert (1991), Griesmer (1991) and Lynch (1991).

References

AEI (2008). *Geoengineering: a revolutionary approach to climate change*. Announcement of a seminar, 3 June 2008. Washington, DC: American Enterprise Institute. Available online at http://www.aei.org/events/geoengineering-a-revolutionary-approach-to-climate-change/ (accessed 12 September 2013).

Atwood, M. (2003). *Arguing against ice cream*. New York Review of Books, 12 June. Available online at http://www.nybooks.com/articles/archives/2003/jun/12/arguing-against-ice-cream/ (accessed 1 June 2014).

Bellamy, R., Chilvers, J., Vaughan, N. E., and Lenton, T. M. (2012). A review of climate geoengineering appraisals. *Wiley Interdisciplinary Reviews: Climate Change*, 3(6), 597–615.

——— (2013). 'Opening up' geoengineering appraisal: multi-criteria mapping of options for tackling climate change. *Global Environmental Change*, 23(5), 926–937.

Brooks, H. (1965). Scientific concepts and cultural change. *Daedalus*, 94(1), 66–83.

Brumfiel, G. (2009). Geoengineering report baffles reporters. *Nature News* blog, 2 September. Available online at http://blogs.nature.com/news/2009/09/geoengineering_does_the_rounds.html (accessed 22 June 2014).

Coale, K. H., Johnson, K. S., Fitzwater, S. E., Gordon, R. M., Tanner, S., *et al*. (1996). A massive phytoplankton bloom induced by an ecosystem-scale iron fertilization experiment in the equatorial Pacific Ocean. *Nature*, 383(6600), 495–501.

Collins, P. (2010). A Royal Society for technology. *Notes & Records of the Royal Society*, 64(Suppl 1), S43–S54.

Crutzen, P. J. (2006). Albedo enhancement by stratospheric sulfur injections: a contribution to resolve a policy dilemma? *Climatic Change*, 77(3/4), 211–220.

Dando, M., Pearson, G., Rózsa, L., Robinson, J. P., and Wheelis, M. (2006). Analysis and implications. In M. Wheelis, L. Rózsa and M. Danda, *Deadly cultures: biological weapons since 1945* (pp. 355–374). Cambridge, MA: Harvard University Press.

Deleuze, G., and Guattari, F. (1988). *A thousand plateaus: capitalism and schizophrenia*. London: Bloomsbury.

ETC Group (2009a). *UK Royal Society on geoengineering: the emperor's new climate?* Press Release, 1 September. Ottawa: ETC Group.

——— (2009b). *The emperor's new climate: geoengineering as 21st century fairytale*. Ottawa: ETC Group. Available online at http://www.etcgroup.org/sites/www.etcgroup.org/files/publication/pdf_file/etcspecialreport_rsgeoeng28aug09.pdf (accessed 1 June 2014).

Fischer, F. (2000). *Citizens, experts, and the environment: the politics of local knowledge*. Durham, NC: Duke University Press.

Fleming, J. R. (2010). *Fixing the sky: the checkered history of weather and climate control*. New York: Columbia University Press.

Fortun, M. (2005). For an ethics of promising, or: a few kind words about James Watson. *New Genetics and Society*, 24(2), 157–174.

Gardiner, S. M. (2011). Some early ethics of geoengineering the climate: a commentary on the values of the Royal Society report. *Environmental Values*, 20(2), 163–188.

Gilbert, S. F. (1991). Epigenetic landscaping: Waddington's use of cell fate bifurcation diagrams. *Biology & Philosophy*, 6(2), 135–154.

Gingrich, N. (2008). Stop the green pig: defeat the Boxer–Warner–Lieberman green pork bill capping American jobs and trading America's future. Letter. *Human Events*, 3 June. Available online at http://www.humanevents.com/2008/06/03/stop-the-green-pig-defeat-the-boxerwarnerlieberman-green-pork-bill-capping-american-jobs-and-trading-americas-future/ (accessed 3 March 2014).

Ginsberg, A., Calvert, J., Schyfter, P., Elfick, A., and Endy, D. (2014). *Synthetic aesthetics: investigating synthetic biology's designs on Nature*. Cambridge, MA: MIT Press.

Gregory, J., and Lock, S. J. (2008). The evolution of 'public understanding of science': public engagement as a tool of science policy in the UK. *Sociology Compass*, 2(4), 1252–1265.

Gribbin, J. (2006). *The fellowship: the story of a revolution*. London: Penguin UK.

Griesemer, J. R. (1991). Must scientific diagrams be eliminable? The case of path analysis. *Biology & Philosophy*, 6(2), 155–180.

Hale, B. (2012). The world that would have been: moral hazard arguments against geoengineering. In C. Preston (ed.), *Reflecting sunlight: the ethics of solar radiation management* (pp. 113–131). Lanham, MD: Rowman & Littlefield.

Harvey, F. (2009). Hopes dashed for geo-engineering solutions. *Financial Times*, 2 September. Available online at http://www.ft.com/cms/s/0/27bee352-9758-11de-83c5-00144feabdc0.html (accessed 1 June 2014).

Hilgartner, S. (2000). *Science on stage: expert advice as public drama*. Stanford, CA: Stanford University Press.

Horton, R. (2003). *Second opinion: doctors, diseases and decisions in modern medicine*. London: Granta Books.

——— (2005). What is the Royal Society for? *The Lancet*, 365(9478), 2174.

House of Commons Innovation, Universities, Science and Skills Committee (2009). *Engineering: turning ideas into reality. Fourth report of session 2008–09, Vol. I, Report, together with formal minutes*. Report HC 50-1. London: Stationery Office.

IMechE (2009). *Geo-engineering: giving us the time to act?* London: Institute of Mechanical Engineers.

Jardine, L. (2003). *The curious life of Robert Hooke: the man who measured London*. London: Harper Perennial.

Jasanoff, S. (1990). *The fifth branch: science advisers as policymakers*. Cambridge, MA: Harvard University Press.

Keith, D. W., Parson, E., and Morgan, M. G. (2010). Research on global sun block needed now. *Nature*, 463(7280), 426–427.

Kruger, T. (Forthcoming). *Dimensions of geoengineering – an analysis of the Royal Society's 'blob' diagram*. Oxford Geoengineering Programme Working Paper. Oxford: Oxford Martin School, University of Oxford.

Latour, B. (2011). Love your monsters. In M. Shellenberger and T. Nordhaus (eds), *Love your monsters: postenvironmentalism and the Anthropocene* (pp. 17–23). Oakland, CA: Breakthrough Institute.

Lawrence, M. G. (2006). The geoengineering dilemma: to speak or not to speak. *Climatic Change*, 77(3), 245–248.

Lentsch, J., and Weingart, P. (eds) (2011). *The politics of scientific advice: institutional design for quality assurance*. Cambridge: Cambridge University Press.

Lindblom, C. (1959). The science of 'muddling through'. *Public Administration Review*, 19(2), 79–88.

Lynch, M. (1991). Science in the age of mechanical reproduction: moral and epistemic relations between diagrams and photographs. *Biology & Philosophy*, 6(2), 205–226.

Macnaghten, P., and Szerszynski, B. (2013). Living the global social experiment: an analysis of public discourse on solar radiation management and its implications for governance. *Global Environmental Change*, 23(2), 465–474.

May, R. M. (2005). Address by Lord May PRS, given at the dedication of the memorial to Dr Robert Hooke at Westminster Abbey on 3 March 2005. *Notes & Records of the Royal Society*, 59(3), 321–323.

Maynard, A. (2009). Geoengineering options: balancing effectiveness and safety. *2020 Science* blog, 1 September. Available online at http://2020science.org/2009/09/01/geoengineering-options-balancing-effectiveness-and-safety/ (accessed 1 June 2014).

Millstone, E., and van Zwanenberg, P. (2001). Politics of expert advice: lessons from the early history of the BSE saga. *Science and Public Policy*, 28(2), 99–112.

NAS (1992). *Policy implications of greenhouse warming: mitigation, adaptation, and the science base*. National Academy of Sciences, Committee on Science, Engineering, and Public Policy. Washington, DC: National Academies Press.

Nordmann, A. (2006). *Noumenal* technology: reflections on the incredible tininess of nano. In J. Schummer and D. Baird (eds), *Nanotechnology challenges: implications for philosophy, ethics and society* (pp. 49–72). Singapore: World Scientific.

NPR/TED Staff (2013). *Can hacking the stratosphere solve climate change?* Available online at http://wap.npr.org/story/209191273 (accessed 6 June 2014).

Olson, R. L. (2012). Soft geoengineering: a gentler approach to addressing climate change. *Environment: Science and Policy for Sustainable Development*, 54(5), 29–39.

Pielke, R. A., Jr (2007). *The honest broker: making sense of science in policy and politics*. Cambridge: Cambridge University Press.

Pielke, R., Prins, G., Rayner, S., and Sarewitz, D. (2007). Climate change 2007: lifting the taboo on adaptation. *Nature*, 445(7128), 597–598.

Porter, T. M. (1996). *Trust in numbers: the pursuit of objectivity in science and public life*. Princeton: Princeton University Press.

Reuters (2009). *World must plan for climate emergency – report*. Reuters, 2 September. Available online at http://www.reuters.com/article/2009/09/02/idUS125187741423._CH_.2400 (accessed 1 June 2014).

Ricke, R., Morgan, M. G., Apt, J., Victor, D., and Steinbruner, J. (2008). *Unilateral geoengineering: non-technical briefing notes for a workshop at the Council on Foreign Relations, Washington DC, May 05, 2008*. Washington, DC: Council on Foreign Relations. Available online at http://www.cfr.org/content/thinktank/GeoEng_041209.pdf (accessed 13 September 2013).

Rip, A. (1986). Controversies as informal technology assessment. *Science Communication*, 8(2), 349–371.

Royal Society (2001). *The role of land carbon sinks in mitigating global climate change*. London: Royal Society.

——— (2008). *Joint Science Academies' statement: climate change adaptation and the transition to a low carbon society*. Available online at http://royalsociety.org/uploadedFiles/Royal_Society_Content/policy/publications/2008/7934.pdf (accessed 12 September 2013).

——— (2009a). *Stop emitting CO_2 or geoengineering could be our only hope*. Press Release, 28 August. London: Royal Society.

——— (2009b). *Geoengineering the climate: science, governance and uncertainty*. London: Royal Society.

RS–RAEng (2004). *Nanoscience and nanotechnologies: opportunities and uncertainties*. London: Royal Society & Royal Academy of Engineering.

Schaffer, S. (2010). Charged atmospheres: Promethean science and the Royal Society. In B. Bryson (ed.), *Seeing further: the story of science and the Royal Society* (pp. 131–156). London: Random House.

Schneider, S. (2010). Confidence, consensus and the uncertainty cops: tackling risk management in climate change. In B. Bryson (ed.), *Seeing further: the story of science and the Royal Society* (pp. 425–444). London: Random House.

Shapin, S. (2011). Good housekeeping. *London Review of Books*, 33(2), 19–21.

Sinitsyna, T. (2008). Ways to tame the climate. *RIA Novosti*, 25 June. Available online at http://en.rian.ru/analysis/20080625/112056346.html (accessed 12 September 2013).

Sprat, T. (1958). *The history of the Royal-Society of London, for the Improving of Natural Knowledge*. London: Martyn and Allestry.

Stilgoe, J. (2012). Experiments in science policy: an autobiographical note. *Minerva*, 50(2), 197–204.

Stilgoe, J., Irwin, A., and Jones, K. (2006). *The received wisdom: opening up expert advice*. London: Demos.

Stirling, A. (2008). 'Opening up' and 'closing down' power, participation, and pluralism in the social appraisal of technology. *Science, Technology & Human Values*, 33(2), 262–294.

——— (2010). Keep it complex. *Nature*, 468(7327), 1029–1031.

Tyndall Centre for Climate Change Research (2004). *Tyndall Centre & Cambridge–MIT Institute Symposium: Macro-Engineering Options for Climate Change Management & Mitigation. Special joint symposium held at the Isaac Newton Institute, Cambridge, UK, 7–9 January 2004*. Available online at http://eprints.soton.ac.uk/171559/1/TC_CMI_Symposium_Report.pdf (accessed 12 September 2013).

UNEP (1992). *Convention on Biological Diversity*. Nairobi: United Nations Environment Programme.

van den Daele, W. (1977). The social construction of science: institutionalisation and definition of positive science in the latter half of the seventeenth century. In E. Mendelsohn, P. Weingart and R. Whitley, *The social production of scientific knowledge* (pp. 27–54). Berlin: Springer.

Wilsdon, J., and Willis, R. (2004). *See-through science: why public engagement needs to move upstream*. London: Demos.

Wynne, B. (2014). Further disorientation in the hall of mirrors. *Public Understanding of Science*, 23(1), 60–70.

5 Open-air experimentation

One of the engineers on the Stratospheric Particle Injection for Climate Engineering (SPICE) project, Hugh Hunt, is also 'keeper of the clock' at Trinity College, Cambridge. Trinity was founded by Henry VIII, and it stands out even from its esteemed Cambridge neighbours for academic and architectural reasons. Four centuries after its founding and three centuries after Sir Isaac Newton taught maths there, the college now claims 32 Nobel Prize winners.

The clock sits three floors up, overlooking Trinity's Great Court quadrangle. It needs winding. Hunt and I climb stairs that get thinner and steeper as they approach the attic room that houses the clock's workings and the remnants of student projects investigating the machinery. I dutifully turn the crank that lifts two of the weights, and the conversation turns to geoengineering.

A 100-year-old clock seems like an odd allegory for geoengineering. Hunt appreciates the clock's 'transparency' – the workings of the technology are immediately apparent, unlike, say, that of an iPhone. Inputs and outputs can be traced from pendulums, through the escapement, gears, compensators and connecting rods and out onto the clock face. Geoengineering, if it ever gets off the ground, would require a hugely complicated sociotechnical system, much of which would be dispersed around the globe and whose effects may be invisible. But, for an engineer, the clock serves to illustrate the gap between current thinking about geoengineering and what it would take to make geoengineering a reality.

I ask whether such a clock could be made now. I'm told that it couldn't. The company that made it – Smith of Derby – are still operating, but the craft skills and tacit knowledge have disappeared. The clock was the product of incremental improvements in understanding, design and manufacturing over hundreds of years. The clock follows the model of the 'double three-legged gravity escapement' that controls Big Ben in London's Houses of Parliament.[1] Hunt mentions with some sadness that the Big Ben clock is so constrained by modern demands of control that it has to all intents and purposes stopped telling the time at all. It has to be adjusted to coincide with an atomic clock, so it is a mere projection of another machine's time.

Surprisingly, given its name, the small world of geoengineering research contains very few engineers. As described in previous chapters, the framing of geoengineering research has prioritised assessment of the implications of the technology rather than exploration of the technical feasibility of solar radiation management (SRM). The SPICE project changed this by putting engineering questions alongside more conventional scientific ones.

As one of the first major geoengineering research projects funded anywhere in the world, the SPICE project had few precedents to follow and few rules of engagement. As Fleming (2010, p. 228) has argued, much geoengineering research had until the early part of the twenty-first century been little more than 'geoscientific speculation'. There are no obvious geoengineering laboratories. Geoengineering research does not happen in any particular place. It draws on the tools of various disciplines, but it takes place at their intersections. The coming together of these disciplines and negotiations about which expertise is most relevant and which questions are most important has happened more publicly than in many new areas of science. The SPICE project has therefore provided a unique opportunity to witness the negotiation of responsible science and innovation (see Stilgoe *et al.* 2013).

The life of SPICE

The SPICE project was born in a 'sandpit' run by three of the UK's seven research councils: the Engineering and Physical Sciences Research Council (EPSRC), the Natural Environment Research Council (NERC) and the Science and Technology Facilities Council (STFC). A sandpit is, according to EPSRC, 'an intensive discussion forum where free thinking is encouraged to delve deep into the problems on the agenda in order to uncover innovative solutions'. These occasions bring together 'a highly multidisciplinary mix of participants . . . to drive lateral thinking and radical approaches to addressing particular research challenges'.[2] The research councils are chartered to 'support science' on behalf of the government. They are officially at arm's length from policymakers, protected from government interference by what David Edgerton (2009) calls the 'invented tradition' of the Haldane principle. But in practice they reflect policy priorities onto the scientific community, just as they negotiate scientists' demands with policymakers.

The sandpit is a reflection of twenty-first-century UK science policy. Over the past two decades, the research councils have come under pressure to compensate for the postwar decline in strategic scientific research in government labs and corporate research and development. So while they remain formally responsible for funding world-class basic research, their informal responsibilities have expanded to include a growing focus on innovation and impact. The long-term economic future of nations is seen to depend on their ability to transition to knowledge economies. Science is seen as a vital source of growth and transformation, but the scientific hegemony of Europe and the USA can no longer be assured. The rise of research in India and China has focussed the attention of policymakers on a 'race to the top' (Lord Sainsbury of Turville 2007; see also NAS *et al.* 2007).

The research councils face growing expectations to deliver more than just scientific research as part of a supply-side policy to address a long-diagnosed gap between the UK's stellar scientific achievements and its mediocre innovation performance. The demand side of science and innovation policy receives relatively little attention. (According to Keith Pavitt, 'dealing with deficiencies in business R&D by making basic research more "relevant" is like pushing a piece of string' [Pavitt 1991, p. 117].) The current councils were created from the old Science and Engineering Research Council in 1994 following the Conservatives' *Realising Our Potential* strategy paper (Chancellor of the Duchy of Lancaster 1993), which echoed a common refrain of UK science policy – that the UK was failing to benefit from its world-class science base. EPSRC was chartered to 'promote and support, by any means, high quality basic, strategic and applied research' and also to contribute 'to the economic competitiveness of Our United Kingdom and the quality of life' (EPSRC 1994). The other councils were given similar responsibilities, but each was allowed to develop its own identity according to the particular constituencies they represented. Depending on the cultures of their research communities, different councils now have very different relationships with government, businesses and members of the public.

The sandpit is offered as a solution to a well-known research funding problem. Conventional peer review is seen as too conservative, too slow and too likely to exacerbate rather than cross boundaries between scientific disciplines. Sandpits have therefore been used in areas such as nanotechnology and synthetic biology that promise both scientific excitement and technological spin-offs while bringing science and engineering to bear on social problems from gun crime to mental illness. According to EPSRC's own publicity, this approach is

> a unique concept in scientific thinking. It breaks down barriers and builds new relationships to create a world of endless possibilities. Intensive 'sandpit' workshops assemble the dynamic range of individuals and skills needed to attack real world problems from every angle. Groups and ideas are formed, reviewed and potentially funded within five days.[3]

Scientists apply to attend the sandpit, but they are sifted to limit the number of established researchers. Creativity is prized over experience, so, in the case of the geoengineering sandpit, few of the attendees had much previous involvement with the issues. At the sandpits, a few dozen participants spend five days in a hotel, 'isolated from everyday distractions and stripped of pre-conceptions'.[4] They are exposed to various devices from the toolkits of corporate teambuilding and brainstorming before being shepherded by mentors into multidisciplinary groups. The researchers are asked to propose, discuss and review new scientific research projects. With an allocated research budget on the table, the aim of the sandpit is to award the money to the two or three best groups and their ideas by the end of the week.

For the scientists receiving this injection of pop psychology, it can be disorientating. Indeed, advocates of the sandpit approach would argue that this was the point. One of the SPICE researchers later reflected on what he called '*five days of game show in Cornwall*': '*We spent two days learning about each other and throwing beanbags at each other and shouting names, no exaggeration, which was a bit excruciating . . . I left feeling emotionally drained. It was an absolute rollercoaster.'*

This experience, which this researcher characterised as being '*disconnected from reality*', was seen by a SPICE colleague as being at the heart of their later troubles. SPICE is a project that has struggled to escape the crucible in which it was formed, the sandpit process. The researchers were new to geoengineering and faced with the challenge of coming up with innovative, blue-sky responses. They were told to emphasise innovation and scientific leadership. They were told to collaborate, but their potential collaborators were limited to the other people in the sandpit. At the same time, the teams that self-organise are all competing for a fixed amount of research funds. One can imagine how, for many areas of science and innovation, such a process would produce novel, innovative, possibly groundbreaking research. For areas such as geoengineering, however, the SPICE scientists later concluded that it was utterly inappropriate. One told me that '*in the space of a week . . . you can't get people to think about it fully enough*'. One of the engineers admitted that, when arriving at the sandpit, '*We knew nothing about climate science and even less about the intricacies of dealing with highly charged social, political, ethical issues*'.

In addition to SPICE, the sandpit produced one other project, the Integrated Assessment of Geoengineering Proposals (IAGP). This project aimed at a broad-based evaluation of the various techniques suggested for geoengineering that involved climate modellers, engineers and social scientists. The singular focus of SPICE was easily justified by four decades of scientific speculation, echoed by the Royal Society, identifying stratospheric particle injection as the 'cheapest' and most 'effective' geoengineering proposal.

Despite the almost-random Brownian motion of the five-day sandpit, it makes sense that the scientist running the SPICE project is a volcanologist. As I described in Chapter 3, thinking about stratospheric particle injection has (too closely, it might be argued) followed research on massive volcanoes, whose eruptions spray particles into the stratosphere. The main collaborators who assembled around the SPICE project were climate modellers, chemists and engineers. In exploring the engineering constraints of stratospheric geoengineering, the SPICE project chose to focus not on aeroplanes but on tethered balloons, a deployment mechanism suggested by Lowell Wood, who urged would-be geoengineers to 'pipe it up; spray it out' (Fleming 2010, p. 256). Wood went on to work with Intellectual Ventures, a Silicon Valley firm started by serial inventor Nathan Myhrvold, to patent the idea, later christened the 'Stratoshield' (Chan *et al.* 2008).

By the time of the sandpit, the Cambridge engineers who would end up in the SPICE project had been working with inventor Peter Davidson on Davidson's own idea for a stratospheric balloon, for which he applied for a patent the week before the sandpit was due to begin (Davidson *et al.* 2011). Davidson is a former industrial chemical engineer who spent some years within the UK government as an adviser on innovation policy before becoming an entrepreneur. Davidson had previously expressed an interest in geoengineering and was asked to be a mentor of the sandpit. The potential conflicts of interest were never properly resolved by the Research Council organisers, although Davidson did highlight his previous collaboration with the engineers and recuse himself from key decisions.

During the sandpit, one of the mentors suggested an outdoor experiment. Despite scepticism about its scientific value from the engineers who would conduct it as well as the wider SPICE team, a proposed kilometre-high model of the tethered balloon joined the other SPICE work packages. One of the SPICE researchers joked that '*we went for one, because it's a round number, and kilometre, because it's a standard unit*'. The reasoning was that even if the outdoor experiment did not reveal anything scientifically dramatic, it would grab public attention. But despite input during the sandpit from a representative of Friends of the Earth, who advised on governance issues, neither the researchers nor the funders fully anticipated the type of public debate generated by the experiment.

Informal technology assessment

The proposed outdoor experiment – SPICE's testbed – became a sort of unintended 'breaching experiment'. Breaching experiments, according to Garfinkel, test the unwritten rules of social life by breaking them, 'making commonplace scenes visible' (Garfinkel 1967, p. 36). Even though the testbed constituted around a tenth of the project budget, it attracted almost all of the public attention given to the SPICE project. The events surrounding SPICE (see Box 5.1) were labelled a 'fiasco' by some (*Nature* 2011; Hulme 2012). The SPICE project was always destined to set a powerful precedent for future geoengineering research. As with any major project in a new area, SPICE was not so much following the rules as writing them.

Box 5.1 Chronology of the SPICE project

- 2009 – EPSRC discuss geoengineering but do not commit funding.
- Sept 2009 – Royal Society publishes its report, *Geoengineering the Climate*, recommending funding for geoengineering research. EPSRC decides to allocate funding with other research councils via a 'sandpit' process.
- October 2009 – Research councils convene a scoping workshop aimed at informing a programme of geoengineering research 'which will allow the UK to make informed and intelligent assessments about the development of climate geoengineering technologies' (EPSRC–NERC–LWEC 2009).
- 12 March 2010 – Peter Davidson files a UK patent application for an 'Atmospheric Delivery System'.

(continued)

(continued)

- 15–19 March 2010 – EPSRC–NERC–STFC hold sandpit on geoengineering research. SPICE is one of two successful proposals. Stage-gate process is recommended for the proposed outdoor testbed.
- August 2010 – NERC releases *Experiment Earth? Report on a Public Dialogue on Geoengineering*, although early findings had been fed into the sandpit.
- 1 October 2010 – SPICE project begins.
- November 2010 – Workshop is held with SPICE and invited experts from social science and beyond to scope the criteria to be used at the stage-gate review.
- 15 June 2011 – SPICE stage-gate review panel meets at Imperial College London. Two criteria are passed; three are 'passed pending'.
- 23 June 2011 – Patent application is published.
- 7 September 2011– Phil Macnaghten, chair of the stage-gate panel, writes to EPSRC advising postponement of the testbed until the three outstanding criteria are addressed and evaluated by the panel.
- 14 September 2011 – SPICE hold press conference at the British Science Festival, at which the proposed testbed is publicly announced, with the encouragement of EPSRC. First wave of media coverage occurs.
- 23 September 2011 – SPICE tell EPSRC about the patent application and a likely letter of objection from non-governmental organisations (NGOs). Research council meets with SPICE researchers, Phil Macnaghten and Richard Owen to discuss options for the testbed. Decision is made to postpone the testbed.
- 26 September 2011 – EPSRC announce the decision to delay testbed; letter from NGOs to David Delpy (EPSRC CEO) is copied to UK government ministers and others.
- 26 April 2012 – EPSRC meets with Matt Watson and Richard Owen. Matt Watson decides not to proceed with the testbed.
- March 2015 – SPICE project is to complete its research.

The science writer Jeff Goodell had already predicted that such experiments would be politically problematic, although his own framing was transparent:

> It wouldn't matter that the experiment would be, by any objective standards, as natural as an organically grown carrot . . . In a sense the facts wouldn't matter, because to those people who are fundamentally opposed

> to geoengineering, deliberately messing with the climate is morally and ethically wrong, no matter what the scale, no matter what the impact.
>
> (Goodell 2010, p. 187)

Stakeholders were acutely aware that before governance of geoengineering research or deployment had been decided upon, the potential for a big project to set important precedents for de facto governance had to be taken seriously. In an area that was relatively ungoverned, the way the experiment was designed and communicated came to be hugely important.

In the sense that the idea of the SPICE testbed was disruptive and generated 'a minor public controversy' (Hulme 2014, p. 61), it represented an opportunity for what Arie Rip calls 'informal technology assessment' (Rip 1986). Controversies are often assumed to represent some sort of failure, but if we consider the myriad uncertainties surrounding emerging technologies, they are a vital way for society to make sense of technology. Following Rip's argument, the questions and answers that fall out of such public debates are less controlled than conventional expert-led technology assessment and so may be a more reliable guide to the public credibility of emerging technologies. Controversies provide an early, public airing of the conflicts and terms of contestation that are likely to define a technology's emergence. They challenge taken-for-granted assumptions about science and technology (Brante 1993; B. Martin and Richards 1995). They are opportunities for innovators to slow down, rethink and gain new awareness (Whatmore 2009). Steve Epstein, in his case study of AIDS activists, has argued that 'debates *within* science are simultaneously debates *about* science and how it should be done – or who should be doing it' (Epstein 1996, p. 3, original emphasis). The public nature of the SPICE project meant that such discussions became more visible, allowing for what Nerlich and Jaspal (2012, p. 132) call 'frame shifting'. While society might benefit from controversies that are not just inevitable but healthy, the scientists caught up in a controversy can experience some discomfort. After their project had been funded, the SPICE scientists were informed that they would have to direct at least part of the efforts towards a new, experimental form of governance constructed by the research councils.

EPSRC, as the lead funder of SPICE, were advised by one of the sandpit mentors that the testbed might demand heightened scrutiny. Following discussions with social scientists, including my subsequent collaborators Richard Owen and Phil Macnaghten, EPSRC subjected the SPICE project to a stage-gate before the testbed was due to go ahead. Stage-gating is a well-known technique for companies in new product development, where the decision criteria typically relate to technical and market potential. For SPICE, there were five criteria, relating to the dimensions of responsible innovation I described in Chapter 2. These criteria are described in Table 5.1, along with the recommendations of the stage-gate panel.

The first two of these criteria were conventional, demanding assurances of safety and compliance with relevant regulations. Criteria 3, 4 and 5 were more challenging, asking for reflection and deliberation on the context surrounding

Table 5.1 Overview of stage-gate criteria and panel recommendations

Criterion	*Panel recommendation*	*Comment from the Research Councils (abridged)*
1 Risks identified, managed and deemed acceptable	Pass	No further information required
2 Compliant with relevant regulations	Pass	No further information required
3 Clear communication of the nature and purpose of the project	Pass pending	Additional work is required: (1) a communications strategy; (2) a commitment to two-way communication; and (3) a 'sticky questions' briefing
4 Applications and impacts described and mechanisms put in place to review these	Pass pending	Additional work is required: (1) more information on the envisaged milestones and associated questions that will need to be addressed before deployment of the testbed; (2) a literature review of risks, uncertainties and opportunities of solar radiation management including social and ethical dimensions
5 Mechanisms identified to understand public and stakeholder views	Pass pending	Additional work is required: (1) stakeholder mapping exercise; (2) engagement with stakeholders; and (3) ensuring that key stakeholders are aware of the testbed

the SPICE project. The stage-gate process was appended to the SPICE project after it had been funded, demanding supererogatory engagement from the SPICE team. The SPICE researchers conducted substantial extra work on their own – unfunded by the research councils – exploring the societal context of emerging technologies, as well as working with me with some small additional funding. This process, the stage-gate and its relationship with ideas of responsible innovation are more fully described in a paper by Stilgoe *et al.* (2013). As an experiment in governance it provided an opportunity for collective learning, as well as a forum for the discussion of issues that were starting to be aired in public.

Publicity and postponement

The SPICE testbed was formally announced at the British Festival of Science in Bradford in September 2011. A cycle of discussion and editing had toned down an initial press release that emphasised the novelty and excitement of geoengineering technologies. But by the time of the press conference, an early news piece in

The Guardian had drawn a solid line connecting the experiment to the proposed full-scale technology. Under the headline 'Want to Mimic a Volcano to Combat Global Warming? Launch a Wembley-Size Balloon', the story announced 'the world's first major "geo-engineering" field-test . . . The ultimate aim is to mimic the cooling effect that volcanoes have when they inject particles into the stratosphere' (Vidal 2011).

John Vidal, environment correspondent at *The Guardian*, had reported two months earlier on a June 2011 meeting of an Intergovernmental Panel on Climate Change (IPCC) working group on geoengineering, focussing on an open letter of opposition organised by ETC Group (2011). The September story similarly included critical quotes from NGO spokespeople, repeating the arguments that stratospheric particle injection was so risky and uncertain as to make research at best pointless and at worst dangerous.

Subsequent coverage, following a press conference with the Science Media Centre at which SPICE researchers discussed their own concerns, was less critical. The assembled UK science correspondents covered the story with the uncritical conventions of science journalism, highlighting the novelty of the project, describing it as 'bizarre', drawing allusions with science fiction and repeating references to an 'artificial volcano'. There was little mention of the stage-gate governance process or public deliberations in which the project had already been involved. But the SPICE researchers were quoted describing the uncertainties and controversies involved and the need to 'stimulate public debate'.[5]

The new public prominence of the SPICE project had attracted the attention of a group of small NGOs similar to those who had complained about the IPCC's recent interest in geoengineering. Their open letter – 'Say No to the "Trojan Hose": No SPICE in Our Skies, Say Environmental Justice Groups' (ETC Group 2011) – was copied to the heads of the research councils, five government ministers and the vice chancellors of the universities involved. The letter rejected the idea that this was mere university science, instead seeing the experiment as a statement of British national interest in geoengineering. The response from Chris Huhne came weeks later, repeating the argument that 'geoengineering research is not a first step to deployment; rather it increases understanding of the issue and allows rational discussion and evidence-based policy to be developed'.[6]

On 26 September, EPSRC announced the decision to postpone the experiment, on the advice of the stage-gate panel, to allow for engagement with stakeholders. The journal *Nature*, which, as well as being a leading scientific publication, provides in its front-half news and comment section a forum for discussion and norm-setting on science in general, reported the postponement of the trial as a response to NGO concerns (*Nature* 2011). In reality, the decision had been made by the research councils before they had received the letter from the NGOs.

Earlier in the summer of 2011, a patent application had been published relating to an 'Atmospheric Delivery System' (Davidson *et al.* 2011). Few people noticed, or drew the link back to the SPICE project, even though the owner of the patent, Peter Davidson, was one of the mentors at the sandpit that awarded funding to SPICE and the other two named inventors, acting as consultants on the invention, were SPICE researchers. This was not the first patent application relating to stratospheric aerosol injection (see Chan *et al.* 2008; Oldham *et al.* 2014), but the other researchers on the SPICE project were surprised to discover that their research was overshadowed by a patent application. This had come to light while the SPICE team were preparing their responses to some imagined 'sticky questions' as part of the work demanded by the stage-gate panel. The realisation that the application had been filed less than a week before the sandpit began prompted EPSRC to ask whether there had been a conflict of interest. The research council appointed two of its members to begin a review. Their report was not fully published, but EPSRC promised to review how money was apportioned in its sandpit processes. EPSRC released a statement saying that

> the sandpit was carried out in accordance with standard EPSRC guidelines and the funding decisions taken at the sandpit were sound . . . as a result of the patent applications, it was possible for an observer to develop a perception that conflicts of interest could exist but found that there was no evidence to suggest any individual used their position to influence the commitment of public funds for their own benefit.
>
> (EPSRC 2012)

If nothing else, this episode clarified the markedly different assumptions made by various actors about what is regarded as problematic in the governance of science. For the engineers, patenting is a normal part of their research. For some ideas, it is seen as a more reliable way to assert their intellectual property than publishing scientific papers. (The SPICE engineers had previously found a draft of one of their papers uploaded without their permission while it was being peer reviewed.) The other SPICE scientists, who were told about the patent application three months after it had been published, following the testbed press conference, were less familiar with the idea of patenting as a normal part of science. Matt Watson was told about the existence of the patent application by Marshall Aerospace, a company acting as consultants for the testbed part of the project. He was still coming to terms with what this might mean a few days later when the SPICE team assembled for one of their regular meetings. Watson later admitted on his blog that the discovery of the patent application had caused him 'significant discomfort' (Watson 2012).

EPSRC did not regard the question of conflicts of interest as sufficiently troubling to either change the process of the sandpit or the people selected to act as mentors. Their lack of attention is a symptom of the wider assumption in research funding that not only sees patents as unproblematic but actively encourages patenting activity. EPSRC's website states that 'it's important to make arrangements for managing the intellectual assets generated by research projects. We encourage the exploitation of the results of all the research we fund'.[7]

Geoengineering was seen as just another research area, proceeding under normal rules. But, as I will discuss in the next two chapters, the disciplines involved have very different norms. While the individual SPICE researchers wrestled with their own ambivalence about geoengineering, the patent application lurked behind them as a fairly unequivocal statement of intent. As one of my interviewees put it, '*It isn't just a balloon in the air any more when the patent ties it explicitly to geoengineering*.'

Reactions to the disclosure of the patent application crossed a wide spectrum. Working with EPSRC and SPICE to consult stakeholders on these issues, I found that some of the more bullish geoengineering researchers outside SPICE, particularly in the USA, could not see what the fuss was about. For them, patenting encourages openness, competition and innovation, which they see as preferable to corporate secrecy. At the other end, there was concern that this might be a small step towards what one interviewee called '*corporate ownership of the control of global temperature*', which would narrow the space for independent testing and governance. Few could imagine a scenario in which global SRM could become a marketable commodity. There would be industrial interests in the components and infrastructure, but this situation would be more mundane than a climate-for-sale. However, most recognised that the uncertainties surrounding geoengineering justified caution rather than what some NGO representatives saw as re-treading the footsteps of biotechnology, in which intellectual property had become a form of de facto governance, generating patent thickets, crowding out particular players and preventing potentially beneficial innovation (Jasanoff 2007).

Patents are increasingly recognised as incapable of containing the politics of new technologies (Hilgartner 2009). They shape innovation in particular directions but are rarely treated as problematic in themselves. The Royal Society (2009) report, for example, mentions neither intellectual property nor patents. If they were considered, the assumption was that they would be irrelevant or that normal rules would suffice. Since the publication of that report, David Keith had been arguing that patents on SRM should be banned (quoted in Mulkern and ClimateWire 2012). Shobita Parthasarathy *et al.* (2012) had previously suggested that geoengineering inventions might be treated as a special case in intellectual property law as nuclear technologies are.

The story of SPICE's entanglement with intellectual property is not the conspiracy aimed at 'privatisation of the global troposphere' imagined by Philip Mirowski (2013, p. 341). The SPICE story does not reveal a deliberate plot, but rather a lack of deliberation on assumptions, a stumbling towards governance-by-default. Many of the concerns of those within and outside SPICE relate to the danger of making rules by precedent and accident. The response of the SPICE researchers demonstrates the positive power of governance-by-precedent, too. Following the realisation about the patent application, the researchers and institutions involved in SPICE changed their collaboration agreements so that intellectual property would not be divided up as usual. The new agreement prevented any of the researchers involved from capturing any intellectual property from SPICE research.

The debate around SPICE tugged on the expectations and norms of science, policy and interest group communities. SPICE provided the substance for a conversation about the morality of geoengineering and geoengineering research.

A letter from the Royal Society's then president, Sir Paul Nurse, argued that geoengineering research was like pharmaceutical testing and should be encouraged on the same grounds (Nurse 2011). A *Nature* feature article in April 2012 drew comparisons between the travails of SPICE and recent research on, variously, mutant influenza, functional magnetic resonance imaging brain scans, laser nuclear fuel enrichment and prenatal genetic diagnosis to suggest the emergence of new ethical dilemmas in science (Brumfiel 2012).

The announcement that the SPICE team had decided not to run the testbed drew another round of media coverage. The SPICE team's decision, and the reasons given, provided an opportunity for more thoughtful commentary on the politics of geoengineering research. The decision prompted an editorial in *Nature* (2012) that called for 'a charter for geoengineering', alongside a news article in the same issue highlighting the patent application associated with the project (Cressey 2012). The editorial used SPICE as evidence of a clear 'problem' with geoengineering research. The 'SPICE fiasco' was described as a 'perfect example of the problems that will persist until geoengineers grasp the nettle of regulation and oversight' (*Nature* 2012, p. 415). The editorial was troubled by 'the lack of an overarching governance framework'. There has been much discussion of the need for 'governance before deployment' (Rayner *et al.* 2013) and 'governance before research' (Hamilton 2013). But it is only through public experimentation that it becomes clear what is at stake. We cannot disentangle research from governance. Both are exploratory. Research involves, at one level, the discovery of what is at stake. As will become clear in this chapter, we need to find ways to connect rather than separate the science and the governance and to realise that in some respects they may be one and the same thing.

'The imaginary made real'

As described in the first chapter, the balloon experiment itself acted as a focus for stakeholder concern, but almost nobody took issue with the direct, proximal effects of the experiment. Everyone – NGOs, scientists, policymakers and others – agreed that the experiment would have negligible local risk or that whatever direct risks would arise would be well understood and mitigated by the researchers. Concerns instead related to the balloon as a symbol of other issues – '*the imaginary made real*', as one stakeholder put it. Interviews with NGOs, scientists and others outside the experiment reveal the tone of concern:

> *'The trial wasn't risky, but it was being done for a reason, and the reason is risky . . . It was clear that this wasn't pure research. The purpose was the problem.'*
>
> *'One question that is too infrequently asked is "why?" It's not a specific concern about the impacts of any one experiment. It's a concern about the implications of those experiments.'*
>
> *'It suddenly becomes real. Not just some scientific fantasy . . . This technology is going somewhere . . . It gives momentum. It breaks a barrier. It's not desktop research any more. It's not speculation using models. It's real, it's tangible, it's*

up there. You can see it. You start to think about what it can do and how it might go wrong. It's much more than a little one-kilometre hose hanging off a balloon in Essex or wherever it's going to be. It has tremendous symbolic significance to all sorts of people in all sorts of ways. It changes the game. It's a milestone . . . Project forward and look back and see the trajectory and you can see why it could be, or could have been, a very notable move along that trajectory. And I think that's what people inside the technology, inside that world, are unable to see.'

Although they would provide different arguments, the SPICE scientists would agree with the conclusion of one NGO representative that '*SPICE is about so much more than a field trial of an injection nozzle*'. For environmental NGOs, geoengineering research was seen as what one called a '*cutting edge geo-politically relevant endeavour*'. NGOs therefore deemed it reasonable to scrutinise research projects in a political light. UK policymakers had, at this point, been relatively silent on geoengineering, although they had begun discussions at a meeting on the Convention on Biological Diversity:

'The British Government hosts a meeting to discuss potential impacts of geoengineering on biodiversity . . . and then suddenly the SPICE thing comes out and you're thinking to yourself, "what signals is the British Government trying to send?" . . . I saw the SPICE project more as a signal than as an engineering experiment.'

'The reason SPICE got jumped on is because it didn't appear to be cognisant of the social and environmental debate going on around it.'

In the SPICE project's presentation to the press the sense of an urgent assessment of a potentially troubling technology was accompanied by a competing narrative. The narrative followed the tone set at the sandpit of an exciting first step towards British scientific and technological leadership. The project itself was funded by the Research Councils, whose agendas are officially disconnected from central government policymaking, a point which government emphasised in its own statements about SPICE. But connections were drawn through the science policy assumptions to which the project and its funders subscribed. One NGO representative saw SPICE '*trying to plant the flag of leadership*' in geoengineering. The publicity invited by and accorded to the SPICE testbed forced an explicit discussion of previously implicit assumptions. As one of the first geoengineering research projects, SPICE provided a seed for the crystallising of concerns about possible and desirable futures, the role of science in bringing these about, and the legitimacy of different perspectives in setting research agendas.

Some of my interviewees – scientists as well as NGOs and others – chose to criticise the science rather than the perceived politics behind the SPICE testbed:

'The wrong science at the wrong time . . . '

'Experiments in geoengineering should focus on understanding the efficacy of geoengineering – how well it will work – or understanding what the risks are and

reducing them . . . And this experiment did neither and so it was misconceived from the start.'

'I see the fun in it, but I'm not sure scientifically it made any sense . . . I mean, it's a symbol for the project.'

Ray Pierrehumbert, a climate scientist who regards stratospheric aerosol injection as 'barking mad', dismissed outdoor experimentation as dangerous:

> The whole idea of geoengineering is so crazy and would lead to such bad consequences, it really is pretty pointless. We already know enough about sulfate albedo engineering to know it would put the world in a really precarious state. Field experiments are really a dangerous step on the way to deployment, and I have a lot of doubts what would actually be learned.
>
> (Quoted in Rotman 2013)

David Keith had previously cautioned that

> taking a few years to have some of the debate happen is healthier than rushing ahead with an experiment. There are lots of experiments you might do which would tell you lots and would themselves have trivial environmental impact: but they have non-trivial implications.
>
> (Quoted in *The Economist* 2010)

He publicly criticised the SPICE experiment during a BBC interview:[8]

> I personally never understood the point of that experiment. That experiment's sole goal is to find a technocratic way to make it a little cheaper to get materials into the stratosphere. And the one problem we *don't* have is that this is too expensive. All the problems with SRM are about who controls it and what the environmental risks are, not how much it costs. It's already cheap. So from my point of view, I thought that was a very misguided way to start experimentation.

On further questioning, it was clear that for some scientists this criticism of the science reflected an instrumental concern that the controversy in which SPICE found itself would jeopardise emerging and fragile geoengineering research agendas: 'M*y fear is the premature efforts could spur some kind of regulatory or funding backlash that would have a negative impact on future research.*'

Scientists typically held more or less well-articulated research agendas with which SPICE was seen to clash. In the sense that all experiments need to hold on to particular certainties in order to probe areas of uncertainty (Rheinberger 1997), the SPICE testbed clashed with a particular set of assumptions about what should be under investigation. The Royal Society (2009) report had concluded that stratospheric aerosol injection was the cheapest and most effective geoengineering proposal currently on the table. For some geoengineering researchers, this answered their questions about affordability and efficacy. For the SPICE team, who justified the testbed with reference to the Royal Society's diagram, it was an invitation to further research.

Geoengineering, as with any emerging interdisciplinary area, brings together different disciplines and different assumptions. As I will explore more in Chapter 7, the confluence of different uncertainties leads to very different ideas about desirable and important research. Some uncertainties, such as the atmospheric and meteorological effects of SRM, are deemed important targets of (indoor) research. Others, such as those surrounding the feasibility and cost of actual engineering, are seen as secondary – an irrelevance or a task for another day. These implicit ideas of 'well-ordered science' (Kitcher 2003), which are far from merely 'scientific', bump up against Cicerone's call to 'proceed as we would for any other scientific problem' (Cicerone 2006, p. 223). Not only is geoengineering more than merely 'scientific', but scientific problems also each possess their own norms and assumptions about how to proceed, which may be contested but are more likely hidden.

Some NGO representatives accepted that technological testing might be valid on the grounds that tests, if taken to breaking point, might rule out particular options. They might see the sense in these views from one of the SPICE engineers:

> *'We have people quite prepared to talk about technology on the assumption that they will work, and yet we don't know any of them will work. And I don't mean "Oh well, if I inject this at this point will it . . . switch off the Indian monsoon, or will it reverse the Jet Stream", No, I'm not talking about that, I'm talking about "Can we do it? Is it actually physically possible?" . . . We've had 10 years' worth of geoengineering research and we know nothing. We really know nothing. And, you know, that's just an embarrassment . . . Let's start doing some research, and then we discover that, actually, a balloon with a pipe is not feasible because the materials are . . . not available then how about we find that out now rather than in twenty years' time?'*

From the perspective of the NGOs, the experiment looked less like a test of a technology and more like a performance, a public display of technological certainty. In this sense, as with other public experiment studies by science and technology studies scholars, the proposed experiment was contested on the grounds that it was not sufficiently experimental (see Collins 1988):

> *'If the concept is to spray water out of a nozzle, what theoretically can go wrong? I mean you can perhaps not get enough pressure up there to spray it out but then you just say we need more pressure . . . it's so simplistic. In a way it was almost . . . a kind of failsafe experiment that was going to be a way in which those interested in geoengineering would be able to convince the media and some sections of the public that it works . . . The idea of testing a kilometre long pipe held by a balloon to pump water vapour sounded laughable because we know that with enough time and money that's a*

> *piece of cake. There's nothing that can go wrong. If we put a man on the moon before I was born, of course we can hang a pipe from a balloon and pump some water.'*

> *'We don't believe that anybody is launching an experiment of this scale to find out it doesn't work . . . I don't buy that in general the reason to do these experiments is to take geoengineering off the table.'*

The SPICE engineers, reflecting on the reception that greeted their proposed testbed, wondered aloud in team meetings whether things would have been different had they not sprayed out the water, or if the balloon had been at 999 metres instead of a kilometre. As part of a serious discussion about the ethics of scaling up technological experimentation, the engineers half-jokingly questioned whether the tiny model balloon-on-a-string that they had built inside their lab was itself controversial. Following the decision to shelve the testbed, they proposed visiting a tethered balloon in Japan, launched for the very different purpose of a robot rope-climbing competition. They wondered if this visit would meet with objections similar to those levelled at their balloon experiment. For the engineers, used to evaluating things in terms of impacts and practicalities, to have been judged so harshly for their intentions was confusing. They asked each other and me what it was about the experiment that so bothered people. Having initially considered the experiment unproblematic, one of the team later said,

> *'I can see the reasons why the one kilometre testbed is controversial, but . . . it's not because of what we want to do in SPICE, it's because of the way people perceive geoengineering, not about SPICE . . . we are rehearsing within this rather small project the debate which has to happen globally. It's fantastic . . . . Yes, I'm disappointed with the delay but it has certainly given the project much more prominence than it might otherwise have had.'*

Another SPICE engineer concluded that '*going ahead with it would set a precedent for outdoor geoengineering tests and there was no governance structure that could manage this*'.

Following the cancellation of the testbed, Hugh Hunt gave a public lecture in which he concluded that the experiment had raised important questions:[9]

> Where does geoengineering stop and research begin? Or is it all mixed in together? This experiment . . . looks as if we're serious about doing geoengineering . . . You get so excited about the technology that you forget . . . that we are screwing up our planet, and is it right to screw it up even more? Who should ask that question? Who should say whether this research should carry on?

The engineers have come to terms with a discussion, described below, that is taking place among the geoengineering community about where to draw the line that defines public concern. Their realisation is that such lines can't be drawn in any simple sense.

Scientific experimentation is rarely as playful or speculative as popular portrayals of science would suggest. Given scarce time and research money, we cannot expect scientists and engineers to exhaust every option. So an experiment implies at least a degree of endorsement. Testing tends to focus on something likely to be successful, with the odds tipped more firmly in this direction if the experiment is public and visible. As Latour (1990) and Schaffer (2007) describe with reference to Archimedes and Guericke, respectively, a well-constructed public experiment carries a political weight far greater than more abstract, more widely applicable mathematical claims. Critics of the SPICE project saw it as a demonstration (in both technical and political senses of that term; Barry 1999) of technological hubris, designed to speak for itself and so change the debate on geoengineering.

Many of the lines of debate that were heard around SPICE were echoed within the project itself. One SPICE researcher admitted that the experiment seemed to be '*opening the gates to something else*'. Another SPICE scientist told me, '*I totally agree with all the concerns that the public had and we hadn't really thought about them and talked about them.*' The team would, later in the project, discuss and engage with such concerns at great length. For the non-engineers in SPICE, the testbed, which one referred to as '*the world's biggest water feature*', seemed to be a step towards a future that they regarded as hugely troubling.

As the SPICE team grew to appreciate their project's entanglement in the politics of geoengineering, they developed nuanced understandings of the dangers of lock-in and the slipperiness of the slope towards deployment. Unlike other areas of emerging science, such as nanotechnology or synthetic biology, in which researchers are typically optimistic about the sociotechnical futures being imagined, the SPICE team were ambivalent, if not terrified, about a geoengineered future.

Governing experimental systems

The reframing of the SPICE testbed represents a radical expansion of the experimental system. The experiment was initially viewed by the researchers and the funders as a test of a technology. The system was seen as a purely technical one. The implications and risks seen as relevant were direct ones – the risks to participants, local communities and bystanders. The relevant public was seen as those people in the immediate vicinity of the airfield on which it would take place.

The reframing of the SPICE experiment in public led to a rethinking of the relevant public and the relevant issues. The experimental system extended beyond the immediate apparatus to encompass an imagined technological trajectory at which the balloon appeared to point. The relevant environmental risks therefore were not just the direct ones but the indirect and uncertain risks of large-scale geoengineering. The SPICE experiment was not the first outdoor geoengineering experiment, nor was it the most controversial. But it was the first outdoor experiment to not only label itself unashamedly as a geoengineering experiment, but also to engage with the wider questions in which it was entangled. It therefore shouldered additional responsibility for setting particular precedents and engaging with the 'second order risks' (Wynne 1992) of subsequent developments. In engineering parlance the system, imagined at first to be closed, came to be seen as open, not to the flow of energy and mass, but to flows of knowledge, interests and opinions.

In the sense that SPICE invited public attention, it consciously expanded its public, although there was no obvious answer to the question of when this expansion should stop, given that a successful geoengineering technology would in principle make the entire global population stakeholders. Recognising the experimentality of this public engagement, some SPICE researchers collaborated with their counterparts at the IAGP, which included social scientists with experience of deliberative engagement. They conducted a small public engagement exercise at which the SPICE experiment was used as a stimulus. The concerns that were drawn out reflected well the debate that was taking place concurrently among stakeholders. The experiment was seen as posing some new local risks, but most worries related to the idea of geoengineering itself and the role that SPICE would play in bringing this about (see Pidgeon *et al.* 2013 for a report of this work).

Experiments involving other forms of geoengineering had brought similar questions to light. In 1999, a proposed experiment in carbon sequestration off the coast of Hawaii, which most scientists agreed was environmentally inconsequential, attracted fierce opposition from local communities and international environmental NGOs, who contested not just the safety of the experiment but also the politics of technofixing the problem of carbon dioxide emissions. The experiment was moved from Hawaii to Norway but met similar resistance (de Figueiredo *et al.* 2003).

More relevant to SRM, the Eastern Pacific Emitted Aerosol Cloud Experiment (E-PEACE) in 2011 measured the reflective effects of various particles using a ship, an old Army smoke generator and an aeroplane. A scientific paper published by the researchers framed the results in terms of 'gaps in fundamental understanding of cloud processes', asking 'how can the understanding of cloud responses to increased aerosol levels be represented in theories and models of the climate system?' (Russell *et al.* 2013). The experiment was not initially described as relevant to geoengineering. Rather, it sought to investigate cloud formation, one of the more important but poorly understood parts of the climate. However, a concurrent paper from the project's leader, published in a National Academy of Engineering magazine, linked the study explicitly to geoengineering, in particular the cloud brightening approach put forward by Latham and Salter (Latham 2002; Salter *et al.* 2008). This paper concluded that 'the E-PEACE results provide a proof of concept that cloud brightening to reduce global mean warming is possible, with existing, decades-old technology, for some cloud conditions' (Russell 2012, p. 14).

Blurring the lines between science and geoengineering still further was a 2012 experiment by the Haida Salmon Restoration Corporation (HSRC) off the coast of northern Canada. More than 100 tonnes of iron sulphate was dumped from a ship far enough off the coast to be in international waters, which, like the stratosphere, represent an under-governed space. This ocean iron-fertilisation

(OIF) project was justified by the Haida Nation community as an attempt to explore whether salmon populations could be brought back to previous levels by improving the fish's food stocks. However, at various times, additional justifications were offered for the iron dumping. While community leaders had imagined the experiment to be only of local interest, suspicion grew that there were grander motivations in play. Campaigners, including ETC Group, drew the connection to geoengineering and questioned the legality of the action. *The Guardian* published a story headlined 'World's Biggest Geoengineering Experiment "Violates" UN Rules' (see Buck 2014).

Part of the reason for controversy in the HSRC case was the involvement of Russ George, a professed 'garage experimenter' with interests in cold fusion. George had previously conducted some amateurish OIF experiments (using a boat loaned by the singer Neil Young). George and his company Planktos (motto: 'Save the world, make a little money on the side') were interested in using OIF to sell carbon credits – an idea that subsequently found its way into the novel *Solar*, Ian McEwan's (2010) satire of eco-capitalist science.

OIF had interested geoengineering researchers since oceanographer John Martin first described the 'iron hypothesis' in 1988 (J. H. Martin and Fitzwater 1988). His tongue-in-cheek claim – 'give me half a tanker of iron, and I'll give you an ice age' – provides an Archimedean mantra for the Anthropocene. In principle, OIF is a high-leverage technology, escaping the engineering constraints of other carbon dioxide removal (CDR) proposals. Each atom of iron could, by fertilising algal blooms, remove thousands of atoms of carbon. Testing of the iron hypothesis rapidly took on a commercial as well as scientific flavour, with scientists in the early 2000s discussing the potential for selling credits on newly created post-Kyoto carbon markets (e.g. Markels and Barber 2001). In practice, fertilising oceans seems to be more problematic than the theory suggests. Experiments carried out have provided mixed results, which may be explained by experimental design, local ecology or fundamental unpredictabilities in ocean systems. While some researchers press ahead (e.g. Smetacek *et al.* 2012), others have argued that the scheme is a non-starter (Strong *et al.* 2009).

These OIF experiments have been variously challenged by environmental campaigners on the grounds that they are illegal, unpredictable or undesirable. Scientists have responded that as with SPICE, the direct environmental implications are likely to be minuscule. Ken Caldeira, discussing the HSRC experiment, argued that 'the trawlers and fishing boats that operate every single day do much more harm to ecosystems than what Russ George did'. For Caldeira, the experiment is about 'slippery slopes and precedents and all this, and not really the action in and of itself' (quoted in Kounaves 2013). International law has moved swiftly to ban this sort of dumping at sea except for 'legitimate scientific research' (Strong *et al.* 2009). Conventional scientists have distanced themselves from George, questioning the extent to which his experiments are about science or simply about making money. George admitted to one journalist, when questioned about the absence of research papers, that 'it's really more of a business experiment than a scientific experiment' (quoted in *Living on Earth* n.d.).

As with SPICE, the HSRC experiment was contested less for its risks than for its wider meaning (Kounaves 2013). It is convenient to dismiss George as a 'vigilante' or 'rogue' geoengineer.[10] Indeed, there is a common tendency to use 'bad apples' to close down discussions of governance, even though there is little agreement on the line between 'good' and 'bad' science (Fanelli 2011).

Drawing lines

These experiments, and the breadth of debate around geoengineering, have prompted some scientists to try to reassert control over research agendas. Geoengineering research can only go so far with models and observations. At some point, some researchers argue, experiments that are 'perturbative' (that is, having an intended effect on the atmosphere; Russell *et al.* 2012) will need to be performed *in* the environment, *on* the environment. Conventional climate scientists are also keen to ensure that the experiments they want to perform on the climate, in order to learn things that modelled projections or straightforward observations cannot tell them, are allowed to go ahead. Their fear is that research agendas are disrupted if such experiments attract the spotlight that was shone on SPICE. The debate around SPICE seemed to pose a clear challenge to scientific autonomy by suggesting that non-scientists may have legitimate concerns about experiments that were not directly risky. Despite the decision being made by the SPICE team themselves, some scientists perceived the testbed cancellation as being the result of public opposition (e.g. Olson 2012). Edward Parson and David Keith published a prominent article in *Science*: 'End the Deadlock on Governance of Geoengineering Research'. Although they don't mention SPICE, it is clear that the project casts a shadow over their thinking. In the tradition of 'social contract' (Guston and Keniston 1994) thinking, they concede that trading 'a modest regulatory burden' for a degree of autonomy is worthwhile in 'allowing small scale research to proceed' (Parson and Keith 2013, p. 1279).

Parson and Keith identify two dominant positions that have emerged. The first sees geoengineering research in terms of a slippery slope towards deployment, justifying a complete moratorium on research. The second is that scientific freedom should dominate and that in the absence of any risk or ethical rights violations, geoengineering research should be governed as any other area of science. Parson and Keith attempt to chart a course between these. They tack towards the former in recognising the need to give scientists 'guidance on the design of socially acceptable research' and address 'legitimate public concern about reckless interventions or a thoughtless slide from small research to planetary manipulation'. They also caution against dismissing public concerns as 'unscientific'. They acknowledge the value of some sort of moratorium and agree that 'controversies should be expected' (Parson and Keith 2013, p. 1279). But their argument then veers towards the latter position in suggesting a threshold for this moratorium.

The lines they draw are at particular levels of 'radiative forcing perturbation' (ΔRF). They suggest that experiments should be banned above a particular level, $\Delta RF > \sim 10^{-2}\ W\ m^{-2}$ (i.e. the change in radiation forcing is greater than approximately 0.01 watts per square metre), at which the effects would be detectable. They also suggest a threshold far below this, $\Delta RF > \sim 10^{-4}$ W m^{-2} (i.e. the change in radiation forcing is greater than approximately 0.0001 watts per square metre), under which experiments should be allowed to proceed. The logic runs that experiments that would be performed in the foreseeable future are likely to have less direct impact on the environment than things that are already taking place. OIF experiments, it is argued, introduce less pollution into the oceans than sewage outfalls. And SRM experiments, even if they are perturbative, are still negligible compared to the cocktail of accidental perturbations constantly imposed on the environment.

Following the Royal Society's recommendation of 'the establishment of a *de minimis* standard for regulation of research' (Royal Society 2009, p. xii), the Solar Radiation Management Governance Initiative (SRMGI) became a forum for the circulation of discussion about when governance should happen: Now or later? Above a particular perturbative threshold? When experiments go outdoors? At a particular scale? (How would we define 'scale'? Length of impact? Size of inputs? Size of impact?) Some of the SRMGI scientists argued that it was impossible to find this line (like '*drawing the edge of a cloud*', according to one), while others insisted that lines should be drawn around an '*allowed zone*' nevertheless. A report from the US Congressional Research Service talked about this in terms of a 'threshold for oversight' (Bracmort and Lattanzio 2013).

Victor *et al.* (2013, p. 3) agree that 'the key is to draw a sharp line between studies that are small enough to avoid any noticeable or durable impact on the climate or weather and those that are larger and, accordingly, carry larger risks' (see also Parson and Ernst 2013). Many geoengineering researchers have attempted to draw a line between CDR and SRM, suggesting that the two are so different in terms of technology, risk and governance as to constitute completely separate issues (see Chapter 3). The Royal Society (2009) Working Group on Geoengineering, prompted by Steve Rayner, made the point that we could equally cut up geoengineering proposals in terms of whether they are self-contained engineering, such as direct air-capture machines, or the leveraging of ecosystems, which would include OIF, as well as stratospheric particle injection. The CDR and SRM communities of interest cannot be so easily delineated. Prominent geoengineering researchers, such as Ken Caldeira and David Keith, actively research either side.

Ethicists, social scientists, lawyers and others in the orbit of geoengineering have joined the project of line drawing. Morrow *et al.* (2009, p. 1) discuss ethical guidelines that might apply 'in the event that CE [climate engineering] research progresses beyond computer modelling'. Alan Robock (2012) has attempted to construct a logic for prohibition of outdoor experiments – with the corollary being that indoor research should be allowed to proceed as normal. James Fleming jokingly refers to research remaining 'indoors, between consenting adults' and argues that 'what needs to be aired out are the underlying assumptions' (Fleming 2010, p. 257).

These proposals are not merely technical line drawing. The attempt to delineate good research from bad, valuable experiments from controversial ones, is also a construction of what Parson and Keith call 'legitimate public concerns' (see Stilgoe 2007). It betrays a particular, impoverished idea of what governance is and how it relates to scientific research. Governance is seen as regulation or even prohibition. It is something that is done *to* scientists rather than *by* them and is therefore to be avoided. This idea, more American than European, is built on a particular version of scientific freedom, or the 'right to research' (Brown and Guston 2009). To use Isaiah Berlin's (1958) phrase, it is 'negative liberty', freedom from interference. The drawing of thresholds is a formalisation of this view. One scientist expressed it like this:

> *'I would start from a presumption that if people are not doing direct harm to other people or property, they should be allowed to do things, and that's a basic sort of libertarian sense of freedom of action and freedom of scientific research and technical investigation. But I don't think this is absolute, because we could imagine a benign experiment where that experiment is one step or two steps away from creating a virus that will kill everybody in the world.'*

As Brown and Guston (2009) describe, there is another way of constituting scientific freedom, which is to embed science within society rather than seeking to extricate it. Science is always and everywhere governed, most obviously by scientists themselves. We might ask why this governance, including the norms and cultures of science, should not also be open to wider discussion.

This is a not an argument for less experimentation. Indeed, it becomes an argument for broadening the scope of experiments. Historians of science have described how early modern scientists tried to negotiate experiments as safe spaces for dissent (Shapin and Schaffer 1985). The discussions taking place within and around geoengineering experiments revealed similar concerns. Some SPICE researchers appreciated the expansion of debate around their experiment and recognised the challenges it posed to conventional notions of governance:

> *'People want to draw a bright line . . . and say everything above it is legitimate and everything below it is dangerous and requires governance. But that* [laughs] *. . . that attitude undermines everything that SPICE is trying to figure out, everything that SPICE has been challenged to do in terms of looking towards the far field, thinking about things like lock-in.'*

For scientists, the challenge is to recognise and respond to this democratisation of experimentation.

Governing intent

Arguments about the governance of technology often hinge on the issue of dual use. Powerful technologies and research fields, even if imagined as neutral, can have 'good' or 'bad' uses. Knowledge about, and technologies for, nuclear fission can be used to build nuclear power plants or nuclear bombs. Certain organophosphates can be used to produce pesticides or chemical weapons, such as the nerve agent VX (McLeish and Balmer 2012). In such areas, the control of technological artefacts alone would be all but impossible, so governance also pays attention to motivations. The Biological Weapons Convention (UNDDA 1975), for example, governs intent. Whether or not scientists' possession of anthrax is seen as problematic depends on what they intend to do with it: weaponise it or attenuate it to develop a vaccine (Tucker 2012). But the governance of intent does not sit easily in science and innovation, whose governance regimes tend to focus on outcomes. Clarity of intent invites responsibilities that science is not well equipped to deal with (Douglas 2009).

With emerging biotechnologies we cannot be certain about the artefacts that will have the most potential for abuse, but similar worries are articulated. Prominent scientists such as Bill Joy (2000) and Martin Rees (2003) and campaign groups (ETC Group 2007) have described the possibility of technological catastrophe and 'existential risk' through acts of bio*terror* or bio*error*. Similar concerns have begun to bleed into geoengineering discussions. The OIF experiments described above prompted the London Convention and Protocol (IMO 2006), which seeks to control pollution at sea, to prohibit OIF unless it is deemed 'legitimate scientific research'. The identification of legitimacy here is clearly not straightforward (Buck 2014) – as discussed above, Russ George has justified his experiments, before the fact and after, in various ways, including as legitimate science. But the inclusion of intent as a concern is notable. This case and the SPICE testbed discussed in this chapter tell us that experiments cannot be disentangled from their imagined futures. With highly contentious science, the motivation for experimentation matters as much as, if not more than, the apparatus. Indeed, it may be only through the design and publicity of such experiments that imagined futures are revealed. The SPICE testbed, which seemed to some to be a step towards a rather troubling future, prompted a debate that would not have taken place if research had stayed in the lab.

With geoengineering, we should be as concerned about use as we are about abuse.[11] An important lesson for governance is that, on the whole, people were not worried about the SPICE testbed going wrong; they were worried that it would go right. Geoengineering is unusual among emerging technologies in that it is defined by intent. Although some have insisted that the environmental impacts that mark the Anthropocene amount to 'accidental geoengineering', most serious researchers would agree with David Keith (2000), who defines geoengineering as being about clear planetary-scale intent.

The term 'geoengineering' collects a disparate set of proposals and technologies under the shared motivation of intentionally changing the global climate. This is not to say that all geoengineering researchers agree on the reasons to geoengineer or the purposes of research. Some talk about climate remediation (BPC 2011); others talk about creating new climates. Turning to policy, some researchers argue about whether the aim is to buy time for climate change mitigation, 'shave the peak' from the worst of our global warming projections, protect certain areas such as the Arctic or reduce average temperatures. Many researchers are unwilling to entertain these sorts of discussions, preferring instead to justify their work in terms of merely understanding natural processes using new tools, assessing the risks of technological deployment. In the search for legitimacy, they are as likely to justify their research with reference to others' interests. They are looking at geoengineering 'because it's there', although, as I have described in this book, it only is if we want it to be.

Despite this tangle of motives, however, geoengineering invites important discussions about what science is for. There have been debates about what we might call geoterror and geoerror, but the more important governance discussion is concerned with what happens when the imagined technology works as intended. Too often, technologies emerge without explicit discussion of their purposes. Nanotechnology, for example, has been called 'an amorphous technology that promises to change everything, but nothing in particular' (Nordmann 2004, p. 112). This slipperiness makes vital discussions about responsibility, research and development harder to open up.

An important lesson from studies of responsible innovation is that we do not have to identify conspiracies. Bad technological trajectories may well be paved with good intentions, just as technologies created for bad reasons might be repurposed in the idealised swords-into-ploughshares way.[12] The future of geoengineering is chronically uncertain. If there is a slippery slope, its gradient and direction are not set. Some geoengineering researchers have proposed Manhattan Project–style programmes of technological development (Michaelson 1998; Lempert and Prosnitz 2011; Davidson *et al.* 2012). But, given that the technologies currently in mind for SRM are rather low-tech, we are more likely to see a bricolage[13] of borrowed, adapted and hacked technologies. The SPICE engineers freely talk about the possibilities for using high-altitude balloons in myriad ways, including mobile communications and surveillance. Once such things exist, the suggestion is that it would be easy to repurpose them for particle injection. Just as the serendipities of science resist strong definition according to purpose, so technologies can, too. Even a saw, as Tim Ingold (2011) describes, does not just cut wood: in skilful hands, it can become a (rather unusual) musical instrument. Technologies have unintended as well as intended implications. This is the sense in which designer David Pye (*The Nature of Design*, 1964, quoted in Harford 2011) argues that 'nothing we design or make ever really works'. This does not absolve innovators of responsibility. Technologies are often designed with purposes in mind, just as science is often conducted in the context of application (Gibbons *et al.* 1994). But we need to be sophisticated about the trajectories along which technologies are imagined.

There is a version of the SPICE story that regards the project as a failure of governance. Certainly there were governance mistakes made, mostly by the Research Councils who funded the project. But these mistakes, such as the failure to take seriously issues of conflicts of interest and intellectual property, can be put down to a misunderstanding of what was at stake. SPICE, as a wider social experiment, helped to clarify the stakes of geoengineering research. The SPICE scientists made a decision not to run the experiment that had attracted too much attention, and they used the opportunity to rethink their project and its governance. This has set positive precedents on questions of intellectual property and interdisciplinary research, although the scientists have taken on more than their fair share of responsibility, leaving their funders, who should have joined them, free to claim autonomy. Nevertheless, the SPICE project revealed to its team and its observers that governance was not just something done to science, but something that was done by scientists, too.

To discussions of deviance in science and innovation – terror and error – we should therefore add *emergence* as a far more powerful force. This has profound implications for governance and research, but these implications have the potential to be hugely constructive. We need to get beyond the paralysis that comes from a fear of misuse and instead confront the realities of intent, development and use and how these feed back into research agendas, explicit or implicit.

Notes

1 See http://trin-hosts.trin.cam.ac.uk/clock/?menu_option=history (accessed 1 December 2014).
2 From http://www.epsrc.ac.uk/funding/howtoapply/routes/network/ideas/whatisasandpit/ (accessed 1 December 2014).
3 See http://www.epsrc.ac.uk/newsevents/pubs/welcome-to-the-ideas-factory-home-of-innovation-since-2004/ (accessed 24 October 2013).
4 Ibid.
5 See, for example, Cohen (2011), Connor (2011), Cookson (2011), and UK Press Association (2011).
6 Letter from 'The Rt Hon Chris Huhne MP', Secretary of State for Energy and Climate Change, to Helena Paul, November 2011. Available online at http://www.handsoffmotherearth.org/wp-content/uploads/2011/11/Helena-Paul-1.pdf (accessed 24 July 2014).
7 EPSRC (n.d.). Intellectual assets. Available online at http://www.epsrc.ac.uk/funding/howtoapply/basics/ip/ (accessed 24 October 2013). With thanks to Hugh Hunt for pointing me to this advice.
8 Interviewed on *HARDtalk*, BBC News channel, at 4:30 AM, 14 Nov 2011.
9 Hugh Hunt, lecture at Trinity College Cambridge, 26 February 2013.
10 See, for example, Marshall (2012) and Thomas (2013).
11 For a fuller discussion on these points, see Szerszynski *et al.* (2013).
12 See the chapters in Owen *et al.* (2013), especially that of Grinbaum and Groves (2013).
13 Steve Rayner, Oxford University, personal communication, 2012.

References

Barry, A. (1999). Demonstrations: sites and sights of direct action. *Economy and Society*, 28(1), 75–94.

Berlin, I. (1958). Two concepts of liberty. In I. Berlin (1984), *Four essays on liberty* (pp. 118–172). Oxford: Oxford University Press.

BPC (2011). *Geoengineering: a national strategic plan for research on the potential effectiveness, feasibility, and consequences of climate remediation technologies*. Washington, DC:

Bipartisan Policy Centre Task Force on Climate Remediation Research. Available online at http://www.bipartisanpolicy.org/library/report/task-force-climate-remediation-research (accessed 20 June 2012).

Bracmort, K., and Lattanzio, R. K. (2013). Geoengineering technologies. In K. Bracmort and R. K. Lattanzio, *Geoengineering: governance and technology policy* (pp. 9–19). CRS Report R41371. Washington, DC: Congressional Research Service. Available online at http://www.fas.org/sgp/crs/misc/R41371.pdf (accessed 11 June 2014).

Brante, T. (1993). Reasons for studying scientific and science-based controversies. In T. Brante, S. Fuller and W. Lynch (eds), *Controversial science – from content to contention* (pp. 177–192). Albany: State University of New York Press.

Brown, M. B., and Guston, D. H. (2009). Science, democracy, and the right to research. *Science and Engineering Ethics*, 15(3), 351–366.

Brumfiel, G. (2012). Good science, bad science. *Nature*, 484(7395), 432–434.

Buck, H. J. (2014). *Village science meets global discourse: the Haida Salmon Restoration Corporation's ocean iron fertilization experiment (case study)*. Geoengineering Our Climate? Working Paper. Available online at http://geoengineeringourclimate.com/2014/01/14/village-science-meets-global-discourse-case-study/ (accessed 29 July 2014).

Chan, A. K., Hyde, R. A., Myhrvold, N. P., Tegreene, C. T., and Wood, L. L. (2008). *High altitude atmospheric alteration system and method*. US Patent Application No. 11/788383.

Chancellor of the Duchy of Lancaster (1993). *Realising our potential: a strategy for science, engineering and technology*. Command Paper Cm 2250 presented to Parliament. London: Her Majesty's Stationery Office.

Cicerone, R. J. (2006). Geoengineering: encouraging research and overseeing implementation. *Climatic Change*, 77(3), 221–226.

Cohen, T. (2011). A balloon the size of Wembley and the world's longest hose . . . can this stop global warming? *Daily Mail*, 14 September.

Collins, H. M. (1988). Public experiments and displays of virtuosity: the core-set revisited. *Social Studies of Science*, 18(4), 725–748.

Connor, S. (2011). High hopes for Norfolk's artificial volcano in fight against climate change. *The Independent*, 14 September.

Cookson, C. (2011). Trial aims to hose down warming climate. *Financial Times*, 14 September.

Cressey, D. (2012). Cancelled project spurs debate over geoengineering patents. *Nature*, 485(7399), 429.

Davidson, P., Burgoyne, C., Hunt, H., and Causier, M. (2012). Lifting options for stratospheric aerosol geoengineering: advantages of tethered balloon systems. *Philosophical Transactions of the Royal Society A: Mathematical, Physical & Engineering Sciences*, 370(1974), 4263–4300.

Davidson, P., Hunt, H. E. M., and Burgoyne, C. J. (2011). *Atmospheric delivery system*. Patent Application No. GB 2476518.

de Figueiredo, M. A., Reiner, D. M., and Herzog, H. J. (2003). Ocean carbon sequestration: a case study in public and institutional perceptions. In J. Gale and Y. Kaya (eds), *Proceedings of 6th International Conference on Greenhouse Gas Control Technologies* (GHGT-6), 1–4 October 2002, Kyoto, Japan, Vol. I (pp. 799–804). Amsterdam: Elsevier. Available online at http://sequestration.mit.edu/pdf/ghgt6_paper_134.pdf (accessed 19 December 2013).

Douglas, H. E. (2009). *Science, policy, and the value-free ideal*. Pittsburgh: University of Pittsburgh Press.

Economist, The (2010). Lift-off. *The Economist*, 4 November. Available online at http://www.economist.com/node/17414216 (accessed 19 December 2013).

Edgerton, D. (2009). The 'Haldane Principle' and other invented traditions in science policy. *History & Policy*, 2 July.

EPSRC (1994). *Royal charter – 1994*. Swinden, UK: Engineering and Physical Sciences Research Council. Available online at http://www.epsrc.ac.uk/about/history/royalcharter1994/ (accessed 13 December 2013).

——— (2012). *SPICE update*. News Release, 17 October. Swinden, UK: Engineering and Physical Sciences Research Council. Available online at http://www.epsrc.ac.uk/newsevents/news/spiceupdateoct/ (accessed 13 December 2013).

EPSRC–NERC–LWEC (2009). *Geoengineering Scoping Workshop – Outputs, London, 19 October 2009*. Swinden, UK: Engineering and Physical Sciences Research Council, Natural Environment Research Council, and Living with Environmental Change. Available online at http://www.epsrc.ac.uk/SiteCollectionDocuments/Publications/reports/ReportOfGeoengineeringScopingWorkshop.pdf (accessed 1 February 2013).

Epstein, S. (1996). *Impure science: AIDS, activism, and the politics of knowledge*. Berkeley: University of California Press.

ETC Group (2007). *Extreme genetic engineering: an introduction to synthetic biology*. Ottawa: ETC Group.

——— (2011). *Say no to the 'Trojan hose': no spice in our skies, say environmental justice groups*. News Release, 27 September. Ottawa: ETC Group. Available online at http://www.handsoffmotherearth.org/wp-content/uploads/2011/09/NR-SPICE-270911.pdf (accessed 19 December 2013).

Fanelli, D. (2011). The black, the white and the grey areas: towards an international and interdisciplinary definition of scientific misconduct. In T. Mayer and N. Steneck (eds), *Promoting research integrity on a global basis* (pp. 79–90). Singapore: World Scientific.

Fleming, J. R. (2010). *Fixing the sky: the checkered history of weather and climate control*. New York: Columbia University Press.

Garfinkel, H. (1967). *Studies in ethnomethodology*. Englewood Cliffs, NJ: Prentice Hall.

Gibbons, M., Limoges, C., Nowotny, H., Schwartzman, S., Scott, P., and Trow, M. (1994). *The new production of knowledge: the dynamics of science and research in contemporary societies*. London: Sage.

Goodell, J. (2010). *How to cool the planet: geoengineering and the audacious quest to fix Earth's climate*. Boston: Houghton Mifflin Harcourt.

Grinbaum, A., and Groves, C. (2013). What is 'responsible' about responsible innovation? Understanding the ethical issues. In R. Owen, J. Bessant and M. Heintz (eds), *Responsible innovation: managing the responsible emergence of science and innovation in society* (pp. 119–142). Chichester: Wiley.

Guston, D. H., and Keniston, K. (eds) (1994). *The fragile contract: university science and the federal government*. Cambridge, MA: MIT Press.

Hamilton, C. (2013). *Geoengineering: governance before research please*. Available online at http://clivehamilton.com/geoengineering-governance-before-research-please/ (accessed 7 January 2014).

Harford, T. (2011). *Adapt: why success always starts with failure*. London: Abacus.

Hilgartner, S. (2009). Intellectual property and the politics of emerging technology: inventors, citizens, and powers to shape the future. *Chicago-Kent Law Review*, 84(1), 197–224.

Hulme, M. (2012). Climate change: climate engineering through stratospheric aerosol injection. *Progress in Physical Geography*, 36(5), 694–705.

——— (2014). *Can science fix climate change? A case against climate engineering*. Cambridge: Polity Press.

IMO (2006). *London Convention and Protocol*. London: International Maritime Organization.

Ingold, T. (2011). *Being alive: essays on movement, knowledge and description*. Abingdon: Taylor & Francis.

Jasanoff, S. (2007). *Designs on Nature*. Princeton, Princeton University Press.

Joy, B. (2000). Why the future doesn't need us. *Wired*, 8(04), 1–11. Available online at http://www.wired.com/wired/archive/8.04/joy.html (accessed 7 July 2014).

Keith, D. W. (2000). Geoengineering the climate: history and prospect. *Annual Review of Energy and the Environment*, 25, 245–284.

Kitcher, P. (2003). *Science, truth, and democracy*. New York: Oxford University Press.

Kounaves, S. (2013). *Climate, carbon and controversy: investigating the Haida Salmon Restoration Corporation Ocean Fertilization Project*. Unpublished MSc dissertation, University College London, London.

Latham, J. (2002). Amelioration of global warming by controlled enhancement of the albedo and longevity of low-level maritime clouds. *Atmospheric Science Letters*, 3(2/3/4), 52–58.

Latour, B. (1990). The force and reason of experiment. In H. E. Le Grand (ed.), *Experimental inquiries: historical, philosophical and social studies of experimentation in science* (pp. 49–80). Dordrecht: Kluwer.

Lempert, R. J., and Prosnitz, D. (2011). *Governing geoengineering research: a political and technical vulnerability analysis of potential near-term options*. Santa Monica: RAND Corporation.

Living on Earth (n.d.). Iron fertilization. *Living on Earth: PRI's Environmental News Magazine*. Available online at http://www.loe.org/series/series.html?seriesID=27 (accessed 2 March 2014).

Lord Sainsbury of Turville (2007). *The race to the top: a review of government's science and innovation policies*. London: Her Majesty's Stationery Office.

Markels M., Jr, and Barber, R. T. (2001). *Sequestration of carbon dioxide by ocean fertilization*. Presented at the 1st National Conference on Carbon Sequestration, National Energy Technology Lab, Washington, DC, 14–17 May 2001.

Marshall, M. (2012). Independent geoengineer's ocean field test condemned. *New Scientist*, 216(2887), 4.

Martin, B., and Richards, E. (1995). Scientific knowledge, controversy, and public decision-making. In S. Jasanoff, G. E. Markle, J. C. Petersen and T. Pinch (eds), *Handbook of science and technology studies* (pp. 506–526). Newbury Park, CA: Sage.

Martin, J. H., and Fitzwater, S. E. (1988). Iron deficiency limits phytoplankton growth in the north-east Pacific subarctic. *Nature*, 331(6154), 341–343.

McEwan, I. (2010). *Solar*. London: Jonathan Cape.

McLeish, C., and Balmer, B. (2012). The discovery of the V-series nerve agents. In J. B. Tucker (ed.), *Innovation, dual use, and security: managing the risks of emerging biological and chemical technologies* (pp. 273–288). Cambridge, MA: MIT Press.

Michaelson, J. (1998). Geoengineering: a climate change Manhattan project. *Stanford Environmental Law Journal*, 17(1), 73–140.

Mirowski, P. (2013). *Never let a serious crisis go to waste: how neoliberalism survived the financial meltdown*. London: Verso Books.

Morrow, D. R., Kopp, R. E., and Oppenheimer, M. (2009). Toward ethical norms and institutions for climate engineering research. *Environmental Research Letters*, 4(4), 045106.

Mulkern, A. C., and ClimateWire (2012). Researcher: ban patents on geoengineering technology. *Scientific American*, 18 April. Available online at http://www.scientificamerican.com/article.cfm?id=researcher-ban-patents-on-geoengineering-technology (accessed 12 March 2014).

NAS *et al.* (2007). *Rising above the gathering storm: energizing and employing America for a brighter economic future*. Washington, DC: National Academies Press.

Nature (2011). SPICE on ice. *Nature*, 478(7367), 10.

——— (2012). A charter for geoengineering. Editorial. *Nature*, 485(7399), 415.

NERC (2010). *Experiment Earth? Report on a Public Dialogue on Geoengineering*. Swindon, UK: Natural Environment Research Council.

Nerlich, B., and Jaspal, R. (2012). Metaphors we die by? Geoengineering, metaphors, and the argument from catastrophe. *Metaphor and Symbol*, 27(2), 131–147.

Nordmann, A. (2004). Social imagination for nanotechnology. In *Nanotechnologies: a preliminary risk analysis on the basis of a workshop organized in Brussels on 1–2 March 2004 by the Health and Consumer Protection Directorate General of the European Commission* (pp. 111–113). Brussels: European Communities.

Nurse, P. (2011). I hope we never need geoengineering, but we must research it. Letter. *The Guardian*, 8 September.

Oldham, P., Szerszynski, B., Stilgoe, J., Brown, C., Eacott, B., and Yuille, A. (2014). Mapping the landscape of climate engineering. *Philosophical Transactions of the Royal Society A: Mathematical, Physical & Engineering Sciences*, 372(2013), 20140065.

Olson, R. L. (2012). Soft geoengineering: a gentler approach to addressing climate change. *Environment: Science and Policy for Sustainable Development*, 54(5), 29–39.

Owen, R., Bessant, J., and Heintz, M. (eds) (2013). *Responsible innovation: managing the responsible emergence of science and innovation in society*. Chichester: Wiley.

Parson, E. A., and Ernst, L. N. (2013). International governance of climate engineering. *Theoretical Inquiries in Law*, 14(1), 307–338.

Parson, E. A., and Keith, D. W. (2013). End the deadlock on governance of geoengineering research. *Science*, 339(6125), 1278–1279.

Parthasarathy, S., Avery, C., Hedberg, N., Mannisto, J., and Maguire, M. (2012). *A public good? Geoengineering and intellectual property*. STPP Working Paper No. 10-1. Ann Arbor, MI: Science, Technology, and Public Policy Program, Gerald R. Ford School of Public Policy, University of Michigan.

Pavitt, K. (1991). What makes basic research economically useful? *Research Policy*, 20(2), 109–119.

Pidgeon, N., Parkhill, K., Corner, A., and Vaughan, N. (2013). Deliberating stratospheric aerosols for climate geoengineering and the SPICE project. *Nature Climate Change*, 3(5), 451–457.

Rayner, S., Heyward, C., Kruger, T., Pidgeon, N., Redgwell, C., and Savulescu, J. (2013). The Oxford Principles. *Climatic Change*, 121(3), 499–512.

Rees, M. (2003). *Our final century: will the human race survive the twenty-first century?* London: Heinemann.

Rheinberger, H-J. (1997). *Toward a history of epistemic things: synthesizing proteins in the test tube*. Stanford, CA: Stanford University Press.

Rip, A. (1986). Controversies as informal technology assessment. *Science Communication*, 8(2), 349–371.

Robock, A. (2012). Is geoengineering research ethical? *Sicherheit und Frieden (Security and Peace)*, 30(4), 226–229.

Rotman, D. (2013). A cheap and easy plan to stop global warming. *MIT Technology Review*, 8 February. Available online at http://www.technologyreview.com/

featuredstory/511016/a-cheap-and-easy-plan-to-stop-global-warming/ (accessed 19 December 2013).

Royal Society (2009). *Geoengineering the climate: science, governance and uncertainty*. London: Royal Society.

Russell, L. M. (2012). Offsetting climate change by engineering air pollution to brighten clouds. *Bridge*, 42(4), 10–15.

Russell, L. M., Rasch, P. J., Mace, G. M., Jackson, R. B., Shepherd, J., *et al.* (2012). Ecosystem impacts of geoengineering: a review for developing a science plan. *AMBIO*, 41(4), 350–369.

Russell, L. M., Sorooshian, A., Seinfeld, J. H., Albrecht, B. A., Nenes, A., *et al.* (2013). Eastern Pacific Emitted Aerosol Cloud Experiment. *Bulletin of the American Meteorological Society*, 94(5), 709–729.

Salter S., Sortino, G., and Latham, J. (2008). Sea-going hardware for the cloud albedo method of reversing global warming. *Philosophical Transactions of the Royal Society A: Mathematical, Physical & Engineering Sciences*, 366(1882), 3989–4006.

Schaffer, S. (2007). *Is seeing believing? Why public experiments often fail and sometimes work*. Neale Wheeler Watson Lecture, Nobel Museum, Stockholm.

Shapin, S., and Schaffer, S. (1985). *Leviathan and the air-pump: Hobbes, Boyle, and the experimental life*. Princeton: Princeton University Press.

Smetacek, V., Klaas, C., Strass, V. H., Assmy, P., Montresor, M., *et al.* (2012). Deep carbon export from a Southern Ocean iron-fertilized diatom bloom. *Nature*, 487(7407), 313–319.

Stilgoe, J. (2007). The (co-)production of public uncertainty: UK scientific advice on mobile phone health risks. *Public Understanding of Science*, 16(1), 45–61.

Stilgoe, J., Owen, R., and Macnaghten, P. (2013). Developing a framework for responsible innovation. *Research Policy*, 42(9), 1568–1580.

Strong, A., Chisholm, S., Miller, C., and Cullen, J. (2009). Ocean fertilization: time to move on. *Nature*, 461(7262), 347–348.

Szerszynski, B., Kearnes, M., Macnaghten, P., Owen, R., and Stilgoe, J. (2013). Why solar radiation management geoengineering and democracy won't mix. *Environment and Planning A*, 45(12), 2809–2816.

Thomas, L. J. (2013). Don't dump iron – dump rogue climate schemes. *The Blog, Huffington Post*, 25 May. Available online at http://www.huffingtonpost.ca/l-jim-thomas/geoengineering-russ-george-iron-dumping_b_3332235.html (accessed 19 December 2013).

Tucker, J. B. (ed.) (2012). *Innovation, dual use, and security: managing the risks of emerging biological and chemical technologies*. Cambridge, MA: MIT Press.

UK Press Association (2011). Scientists to create artificial volcano for climate change experiment. *Daily Telegraph*, 14 September.

UNDDA (1975). *Convention on the Prohibition of the Development, Production and Stockpiling of Bacteriological (Biological) and Toxic Weapons and on Their Destruction*. Geneva: United Nations Department for Disarmament Affairs.

Victor, D. G., Morgan, M. G., Apt, J., Steinbruner, J., and Ricke, K. L. (2013). The truth about geoengineering: science fiction and science fact. *Foreign Affairs*, 92(2), Postscript.

Vidal, J. (2011). Want to mimic a volcano to combat global warming? Launch a Wembley-size balloon. *The Guardian*, 1 September.

Watson, M. (2012). Testbed news. *Reluctant Geoengineer* blog, 16 May. Available online at http://thereluctantgeoengineer.blogspot.co.uk/2012/05/testbed-news.html (accessed 13 December 2013).

Whatmore, S. J. (2009). Mapping knowledge controversies: science, democracy and the redistribution of expertise. *Progress in Human Geography*, 33(5), 587–598.

Wynne, B. (1992). Uncertainty and environmental learning: reconceiving science and policy in the preventive paradigm. *Global Environmental Change*, 2(2), 111–127.

6 Making models

The testbed balloon was intended to be an engineers' model: a device for testing and observing how a full-size balloon might behave in the world. Without it, the engineers have been using computer models to explore how strong the pipes and pumps would have to be and to see how a balloon floating upwards to 20 km would behave. They are still asking real-world questions but doing so with virtual tools. The Stratospheric Particle Injection for Climate Engineering (SPICE) project involves making models of various kinds – mental models, computer models and a range of laboratory experiments. Within the team there is an ecosystem scientist working with a plant model to study the effects of diffused sunlight on leaves, chemists modelling the interactions between various particles and the atmosphere, and physicists working on climate modelling. With its blend of understanding and intervening, geoengineering research invites a new set of conversations between models and the real world.

The range of models in the SPICE project and geoengineering research reflects the range of disciplines involved and their various representations of reality. If we take seriously geoengineering as engineering, we should expect modelling of some desired, feared or in some way altered reality. With geoengineering, we are rapidly seeing the turning of existing science tools to new purposes. The statistician George Box famously said that 'all models are wrong; the practical question is how wrong do they have to be to not be useful' (Box and Draper 1987, p. 74). In areas of politicised, controversial science, we should go past just challenging the accuracy and reliability of models to ask questions such as 'accurate for what?' and 'reliable for whom?' We should follow George Box's invitation to connect the rightness and wrongness of models with their utility, especially when, as with geoengineering, these uses look so problematic.

Experimental climates

As I described in Chapter 3, climate models have become a 'virtual laboratory' for scientists in which projections of future climates are labelled 'experiments' and a range of 'forcings' is applied to assess alternative scenarios. The advancement from understanding the climate to understanding a changing climate to understanding a climate that is 'being changed' makes for a relatively small jump to exploring what happens when intentional changes are exerted on the climate. So while geoengineering research and conventional climate science may have radically different political contexts, for practising scientists the difference is one of degree rather than kind. The emergence of geoengineering as a legitimate science-policy discourse creates new options for climate modellers, who can relatively easily add a projection of a geoengineered world to the countless 'what if . . . ' experiments that are run all the time. One climate modeller told me, '*We've been aware that we could potentially just use our model and see what happens [with geoengineering]. You could publish a lot from that.*'

Another modeller explained in more detail the nature of such experiments:

> '*The main purpose is to project climate for the future, but . . . it depends on the scenario that you're setting. The scenarios usually are "okay we will continue to emit greenhouse gases a certain amount and we will continue emitting more or less aerosol precursors" and things like that. So here we add another element to the scenario: what if in addition to this future greenhouse gases and emissions, we could do something on purpose and not as a by-product?*'

Growing computer power has made such *in silico* experiments relatively frictionless. Even as the complexity of climate models increases, transient climate simulations (representing changing greenhouse gas concentrations over centuries) are able to run multiple possible scenarios of changing climates. Such experiments are behind many of the claims made for geoengineering in the twenty-first century, starting with work by Bala and Caldeira (2000) and reinforced by Crutzen (2006) and others.

Some of the less sophisticated models mimic the effects of solar radiation management (SRM) simply by reducing the model's 'solar constant' – or as some of the scientists casually put it, '*turning down the sun*'. This is seen as adequate for representing the effect of imagined space mirrors, but it is often seen as insufficient for representing stratospheric particle injection. The more advanced models can simulate stratospheric geoengineering by adding sulphur dioxide to the atmosphere at particular places and times and seeing what happens.

As geoengineering research starts to become more organised and the attention paid to its findings grows, researchers face challenges that are familiar to climate scientists. They need to avoid scientific duplication, reconcile a diversity of experimental approaches and consider their wider political credibility. One response to these pressures, growing out of similar efforts within climate science, is the Geoengineering Model Intercomparison Project (GeoMIP). It follows the Coupled Model Intercomparison Project, which came together in the 1990s following recognition that the proliferation of models and projections from the world's scientists might be scientifically inefficient, as well as politically ineffective. The work done by these comparison projects is largely about standardisation, norms and rules. The spirit is explicitly inclusive. They aim to generate model 'spreads', rather than show which models are best, but with an implicit idea that competition will lead to improvement. If early geoengineering research can be said to have a home, perhaps it is located at GeoMIP meetings. These are the places where the growing community of geoengineering modellers gather to share scientific findings and discuss how the science might proceed.

GeoMIP's projections have been included in the fifth report of the Intergovernmental Panel on Climate Change (IPCC 2013), which has decided that geoengineering can no longer be overlooked. The IPCC report is careful to assess geoengineering proposals critically, but their inclusion has nevertheless attracted critical attention. An IPCC meeting in Lima, Peru, took place in June 2011. Despite this being much like the many other multidisciplinary geoengineering workshops that have taken place around the world, the IPCC's involvement was seen as worrying, especially given earlier comments on geoengineering from Christiana Figueres, executive secretary of the United Nations Framework Convention on Climate Change, that were seen as supportive.[1] One of the GeoMIP scientists, who sees the IPCC reports as more scientific than political, argues that

> *'I think it's appropriate [for the IPCC] not to not discuss it, because it's a subject that is discussed in the scientific community and the IPCC report is trying to, well, assess the latest state of climate research . . . I think it's better to cover it than to get the criticism "well why are you not talking about this topic, although people are doing research on it? . . . you're trying to hide something." I'm for being open about it.'*

As the GeoMIP scientists discuss their relevance to policy, they decide which experiments to run on their collected models to give a rational, useful picture of geoengineering ideas that are still speculative and at times otherworldly. In attempting to provide useful answers, GeoMIP embeds its own interpretation of the relevant policy questions. GeoMIP focuses on SRM. Now that its constituent models allow for the injection of particles into the stratosphere, its experiments follow, and so reinforce, the dominance of stratospheric particle injection as the geoengineering proposal of concern.

GeoMIP has, in its first round, standardised four geoengineering experiments. The first of these is the least sophisticated: the models' atmospheres have their carbon dioxide instantly quadrupled while the sun is turned down. The aim of the game, and it is sometimes discussed as if it were a game, is to bring the temperature back to normal. Experiment 2 does the same, but with a gradual increase of carbon dioxide concentration over time (one per cent each year). Experiments 3 and 4 are more complicated. They use one of the IPCC's more optimistic projections for future carbon dioxide emissions, which assumes that the world finds a way to stop global warming by 2100. The aim of the experiments is to bring temperature down by adding sulphur dioxide to the atmosphere. In experiment 3, the sulphur dioxide is gradually increased, whereas experiment 4 has a constant amount each year. After the models are run for 50 years of virtual time, the sulphur dioxide is removed to observe the 'termination effect' of rapid global warming (see Kravitz *et al.* 2011). (Early trials of experiment 3 revealed the difficulty of rebalancing the global temperature, so a simpler version was added in which the aerosols were removed and the sun was turned down.)

The framing of these experiments reflects a particular engagement with policy. They seem calculated to be scientifically interesting, rather than realistic. The assumptions of a quadrupling of carbon dioxide in experiment 1 and of an ambitious pathway for the reduction of greenhouse gases in experiments 3 and 4 suggest particular policy choices. A quadrupling of carbon dioxide is an unrealistic scenario in policy terms, but it makes scientific sense; experiments are routinely run using this scenario to come up with a measure of a model's climate sensitivity.

One junior GeoMIP scientist explained the GeoMIP choices as a fear of the moral hazard scenario – the scientists did not want results to be taken as support for geoengineering, either as a realistic proposition in itself or as an alternative to mitigation. The GeoMIP researchers are more interested in working out the quirks of the models than the implications of the experiments. Researchers ask each other '*how well did we do?*' in balancing the Earth's radiation budget. They discuss how many '*runs of the world*' were needed to reach equilibrium. One researcher admitted that their experiment had taken dozens of trial runs to end up at the right place. Tellingly, when asked whether the number of iterations had been reported on in the relevant scientific paper, this researcher replied that it was unimportant.

There is a black comedy to GeoMIP meetings. The laughter partly relates to the minutiae of organising a new collective area and herding different scientists towards agreement on standards, criteria and modelling scenarios. There are in-jokes about getting the models to do their job ('*every time you touch it, it goes "bang" and doesn't work*'), and they talk about the models having '*a life of their own*' (see Lenhard 2007 for more on this point). The same models are described as behaving differently depending on the particular computer server on which they are '*spun up*'. There is a playful aspect to some of the GeoMIP experiments. They are intentionally unreal. But there is also occasional nervous laughter that comes with the recognition that they are running experiments that at times resemble twentieth-century wargames. One scientist asks, part jokingly, whether it might make sense to invent a new unit, '*deaths per watts per square metre*', as a metric for the impacts of geoengineering.

The results of the models, according to one researcher, are '*surprisingly robust . . . in many areas of the world, almost all the models give a very similar response*'. But this internal consistency, to be admired within the small community of geoengineering modellers, is all too easily interpreted as overconfidence outside. In the next chapter, I explore how scientists are navigating these uncertainties and the responsibilities that accompany them.

Experiments with models; experiments on models

Some of the SPICE researchers are part of the GeoMIP effort. The closer one gets to their research, the more one sees the rough edges of their climate models and the intricacies that are largely hidden from discussions in IPCC reports and elsewhere. Asking models to do new things means exposing new flaws. Experiments using models quickly also become experiments on the models.

Despite some pressure from mentors during the sandpit to differentiate SPICE from previous climate science and mark it out as a genuine geoengineering project, a large slice of the SPICE project investigates whether current climate models can adequately represent the known climatic effects of massive volcanic eruptions. Early work suggests that the models do this very badly, which, for a climate modeller, means lifting the lid on the model and trying to work out why. Inside a climate model's code, it should in theory be clear who has added what, for exactly these moments. But in reality it's not so easy to disentangle the reasoning behind the code:

> *'I actually tried to find out a bit of the code and I emailed a few people, there were a few people's names and I couldn't remember who did what and one of them had some log books, but he'd moved to Australia and he was like, "They're in my attic. When I've sorted things out I'll have a look, and then get back in a few months." And he took a look and he got back and he was like, "No, I can't remember what – who did what bit." So yeah, it can be vague as well. Whether it matters? . . . It was the best they had at the time.'*

Some prefer to keep things simple. One of the SPICE scientists described her model with a degree of affection:

> *'One thing I like about the model that I use, is that it's a one-person model, it's a pocket model. So, I have my version of the model that I've worked on for ten years and I'm sure there's bugs in it . . . but I know what's in there at least and I know what it's been tested on and that sort of thing.'*

Most models are collective efforts, even if their constituent parts are still highly personal. One SPICE scientist describes these models as '*huge beasts. No-one knows all of what's in them.*' They are works-in-progress, not machines built according to a blueprint. They are 'tuned' in various ways to provide a better fit with scientists' observations about, for example, the pattern of a monsoon and to suit the particular uses for which they are constructed.

Tuning up

A climate model has a 'dynamic core', embodying the well-known physics of how fluids flow. But the climate system is too chaotic for deterministic physics to give a complete answer. Models therefore include particular parameters, such as for cloud cover, which are harder to represent using the underlying physics. The parameters of these models then have to be 'tuned' to better reflect observations from lab experiments, weather data or more specific, finer grained models. Tuning parameters in this way has become a major part of the climate scientist's craft (Edwards 2001).

Climate models are relied upon to provide authoritative knowledge for policy while also being constantly improved and tweaked. They are both tools and experiments, trying to inform the future while struggling to represent the past and the present. The same data that are used to tune the models can also be used to test them against observations, leading to criticisms, from within and outside the climate science community, of 'double-counting' (Steele and Werndl 2013) or 'circular logic' (Rodhe *et al.* 2000). There is a continual tension between the assumptions of climate physics – the equations that represent physical processes on the Earth and ensure that models conserve mass and energy – and the tuning required to make the model fit observations. Brian Wynne argues for the recognition of 'the open-endedly experimental nature' of climate modelling (Wynne 2010, p. 292).

With the movement of geoengineering research towards the mainstream of climate science, this tension is exacerbated. One climate scientist described to me the initial enthusiasm of climate scientists who turned their tools to geoengineering:

> *'What other people have done in the past, they've just gone, "Yeah, I'll do a geoengineering experiment. This is what happens to our climate," with models that they know don't do a good job . . . [we need to be] honest about what the models can and can't tell us. Some people it suits not to talk about any uncertainty. "This is what's going to happen, if we do it we can improve those things." . . . I imagine they think "well these are the tools we've got, we've got to do something with them" and you can say "well people are using it to do climate predictions and wanting to change the world's attitudes to release carbon dioxide on the basis of them, therefore they must be good enough".'*

For a junior scientist, the incentives to experiment *using* the model can hugely outweigh the desire to experiment *on* the model:

> *'You've got a certain amount of time and everything's at risk and someone's given you a model and you can run an experiment and you can get a paper out of it and you may not even be trying to hide anything but that's just what you're going to do, you're going to run it and you're going to get data because you need data. So you're not going to actively set out to destroy your model [by exploring every flaw], because nothing may come of it.'*

For those modellers in conventional climate science, these choices are made easier by a well-established set of disciplinary norms and objectives. For those interested in geoengineering, the rules are not set. The tuning, bending and patching of climate models is to an important extent dependent on the imagined uses of those models.

Repurposing climate models

Where the development of climate models once followed the rhythms of scientific curiosity and an ongoing need for weather forecasting, more and more the five- to seven-year cycle of the IPCC beats time for the science to follow. Climate models, through the combined efforts of the climate science community to share and standardise, have become astonishingly powerful. A range of different models converge on a robust consensus that our current rate of greenhouse gas emissions will heat up the planet by an average of a few degrees, although there is a blur between two and six degrees Celsius. As climate scientist Reto Knutti describes it, 'For some variables and scales, model projections are remarkably robust and unlikely to be entirely wrong' (quoted in Turney 2013). Climate scientists have, with substantial effort and despite ever-present fringes of dissent, been able to persuade policy and the public that their models tell us about a changing global average temperature. But if climate science has been relatively successful at persuading the world of its One Big Fact with reference to a range of models, how does that change once the purpose of the models changes? Returning to George Box's quote, the models might rapidly become uselessly wrong once the imagined use changes.

Scientists close to the models recognise that each one has different emphases, different quirks and different inaccuracies. These differences reflect different purposes. Modellers can, according to one,

> *'lose track of the fact that models are set up to answer specific questions . . . if you drew a circle around the climate space that you were studying and the model was set up and tested in that little circle, it's often applied outside of that circle.'*

For climate models to usefully contribute to the debate on climate change, they are necessarily built to account for carbon dioxide levels and other variables that are unlike those in the past or present. The way in which a model is built, tested and refined reflects a particular framing of usefulness, as described by Sundberg (2009). One modeller gave me the example of a UK Met Office climate model:

> *'Their focus is on predictability and they've worked exclusively in one area of the atmosphere. So they may have a preference, possibly unwittingly, towards that area of the model being more useful. So . . . they'll have more investment in that one area, whereas another group may consider that chemistry effects are more important. So you can have models based it seems on where some people's preferences may lie. I don't think at any time it's anything bad . . . but you end up with different models with some different properties.'*

Increasing computer power and numbers of scientists will make it possible to contend with some of the complexity of the system they are trying to model and the growing complexity of the models themselves. But how accurate is accurate enough will depend on the particular demands of policymakers and scientists and the uses they imagine for their models.

The turning of climate models to geoengineering is already revealing their technical limitations. Improvements in the horizontal resolution – the number of squares into which the Earth's surface is divided – of climate models have not been matched by improvements in their vertical resolution. For a climate modeller, this may not be a pressing concern. With weather, most of the action happens nearer the surface of the planet. It is only recently that scientists are starting to recognise the meteorological importance of movements from the stratosphere to the troposphere and to improve their models accordingly. For now, it means the models are limited in their ability to calculate the differences we might expect between a cloud of sulphates injected at an altitude of 15 kilometres and one injected at an altitude of 20 or 25 kilometres. For an engineer, this may mean the difference between geoengineering being doable, plausible or impossible. Aeroplanes can relatively easily reach an altitude of 15 kilometres. Above this, the delivery of particles to the stratosphere starts to look much more complicated. One of the SPICE engineers explained his frustration like this:

> *'They [the climate models] are built where the really interesting things . . . are down at sea level . . . Now you go and put a disturbance in at 20km . . . they're not designed for that . . . And it makes a big difference where you want to inject the particles, because if you inject them too low it gets caught up in the turbulence of the tropopause and your particles come out in the rain. You'd like to inject as high as possible. You'd like to go up to 30km or 40km, but you're not going to get a balloon up there, you're not going to get a plane up there, you might get missiles up there, but . . . So it really matters where. The detail matters.'*

Climate models are exquisitely complex. But to an engineer they can seem a hopelessly blunt tool for assessment or control at a planetary scale, like using a child's drawing as the plans for a skyscraper. Levitt and Dubner (2010, p. 181) quote Lowell Wood describing the models as 'enormously crude'. The climate scientists working on geoengineering remain protective of their models. One told me, in response to a clumsily worded question in which I asked about geoengineering models, '*I wouldn't call them climate engineering models, they're* climate *models*'. Another modeller cautioned that

> *'[t]hey [the models] are not at all as complex as the reality. They're useful for some things but I think it's quite dangerous when they can start to be used [for geoengineering] . . . I don't think you could really risk the lives of something like a billion people on a model that has tuned parameters.'*

However, once models start to be used for geoengineering experiments, improvements in their accuracy and resolution are framed by geoengineering needs. This brings a concern that climate scientists will lose control of their models. One scientist reflected on the improvements in modelling that had come from the attempt to better model volcanic eruptions:

> *'Now we have a model we think we can trust better [with better resolution in the stratosphere], we will start to do geoengineering type experiments . . . Would we have done those experiments without the possibility of geoengineering? No,*

we wouldn't. It wouldn't make sense to say, "Well let's inject some sulphate particles" . . . there's no reason to do that, apart from geoengineering . . . I have concerns that there are groups who just do that and science then is being directed by the politics of geoengineering.'

The concern is that climate models might be used not just to predict the implications of a geoengineered world, but also to act as control devices. This presents a number of fundamental problems for the relationship between these models and decision-making. As climate science is relied upon for policy, the future must be regarded as an extrapolation of the present to make it amenable to prediction and risk assessment, and yet the aim of policy or innovation is to change the future. With economic modelling, the financial crisis has reawakened an understanding of the dangers of imagining models in the present as stationary (Orléan 2010). But, perhaps because climate change has come to be seen as beyond the control of global policymakers, connections between action and prediction in climate science are not well developed.

Looking back to the prehistory of climate and geoengineering research, we see a dominant modernist assumption from people like John von Neumann and Joseph Fletcher from RAND of a line connecting improved computing power to improved model resolution, leading to perfect predictability and therefore to perfect control of weather and climate. Climate science has revealed the complexities of modelling, but assumptions of prediction and control continue to inform much geoengineering thinking.

The engineers see things differently. They know that the models are perennially imperfect. They are critical not just of the choices made in the models, but also of the models' inability to offer reliable, validated findings. The SPICE engineers have learned, from their proximity to the climate modellers, how bad the models are at explaining and predicting the impacts of massive volcanic eruptions. They can be tweaked, tuned or '*parameterised*' to model a volcano in hindsight, but to an engineer this is as useful as predicting yesterday's weather. The climate scientists' response is to improve the models. The engineering response is to find ways to validate or even circumvent the models by directly testing, experimenting and observing. One engineer put it like this:

'I just think the climate modellers are, well, not helping by being so definitive in so many ways about, "Oh, we've done this model and this is what happens." And you think, "Well, where are your error bars? Could you run the model a few more times with different assumptions, different initial conditions and see what the spread of results is?" . . . The climate models are not good enough to tell us what things we need to do. I mean, we'll just have to do the experiments . . . What experiments are we going to do to check it's correct? Do we wait around for the next volcano?'

Geoengineering research, from this view, would become not an exercise of modelling and prediction, but a set of experiments in control.

The science of the unknowable

On a return visit to Trinity College, Cambridge, I received a demonstration in control from Hugh Hunt. I was given a mirror and a snooker ball. A small circle was drawn in the middle of the mirror, and I was told to keep the ball in the middle of the circle. After I failed to do this with a series of wobbles and jerks, my teacher rubbed out the small circle and drew a larger one. With the bounds of my control mechanism having been expanded, the task became much easier. Hunt explained that the principle also applied to his beloved clock. The clock is, for its age, astonishingly accurate. A set of tiny weights can be placed on a platform halfway up the pendulum to correct the clock, but in the main, its temperature compensator deals with the changing weather pretty well. The clock's keeper, with the help of Trinity's engineering students, has to set tolerance limits for the control mechanism. If these are too narrow, if the engineers are too demanding, the clock will paradoxically become less accurate. Rather than weaving gently around a desired average, the clock's timekeeping 'bangs off the sides', bouncing from too fast to too slow. If they make the bounds of tolerance wider, say five seconds of variability on either side, the clock is more accurate over time. A wider tolerance for uncertainty produces more accuracy.

If scientists were to put considerations of engineering back into the modelling studies currently underway, what would that look like? The mode of engagement would not be a modernist predict-and-provide, but would instead be cybernetic. Cybernetics is the science of 'exceedingly complex systems' or the 'science of the unknowable' (Pickering 2004). Emerging after World War II, cybernetics sought to investigate how we could engage with and adapt to systems we could never fully comprehend. The sorts of thinking that emerged with reference to computer algorithms, economies, robotics or the human brain might equally apply to the global environment. The metaphor of the planet as a human body (Nerlich and Jaspal 2012) suits engineers who might see themselves as planetary physicians. Pharmaceuticals are 'underdetermined by their chemical structure' (Lakoff 2008, p. 743). To work out what drugs do, we need to test them beyond just having their effects modelled and predicted. We do not need to understand every physiological detail; clinical trials can be used to skate over this uncertainty. As one of the SPICE engineers explained:

> *'The modelling is never going to tell us [what will happen]. Using the analogy, the human body is really complicated. Try designing a control system for a diabetic without doing experiments. You wouldn't do that. You'd get lots of patients in and you'd do lots of tests and you'd try things out, and then you'd do a randomised trial and so on, because the human body's a really big old thing.'*

Given that the effects of a geoengineering technology will be incompletely predictable in a similar way, we can consider what sorts of system might account for this uncertainty.

The thermostat – a switch that activates at a particular temperature – is among the simplest of cybernetic systems. It is a form of homeostat that sidesteps having to predict changes in temperature. But most geoengineering researchers, when they talk about the implications of a 'planetary thermostat' (e.g. Robock 2012), do not consider the systems of control necessary to establish such a thing (see MacMartin *et al.* 2014a for a recent exception). Much speculative research on geoengineering imagines that a thermostat exists or could be created quickly. In reality, a geoengineered world would have a thermostat as an end goal, rather than a starting point. Even though the use of such a thermostat, if one were available, would itself be a social experiment, as Hulme (2014) argues, the prior social experiment would be to construct a system such that it was possible to control temperature. Planetary geoengineering would not mean immediately taking control of the planet. It would instead be an experiment to see what sort of control was possible. The engineers argue the need to take 'control' seriously, even if some think the calculations are premature. One SPICE engineer told me that '*we're not there yet, we haven't got anything to control*'.

Nigel Clark (2013) correctly observes that geoengineering discourse displays a good deal of caution about the possibility of a fix for climate change. In public, researchers acknowledge that 'engineering' of the planet will be more like 'nudging' or 'tweaking' a complex system (Clark 2013, p. 2831). However, even if no scientist publicly compares the planet to an engine, the sorts of experiments that have come to dominate geoengineering research are framed by predictive and deterministic assumptions. Uncertainties may find their way back into the public discussion of these experiments by way of caveats and other mechanisms that are discussed in the next chapter. But they get squeezed out in the reduction to feasible research experiments.

More recently, geoengineering researchers have begun to engage with the idea of geoengineering as an experiment in control, prompted by questions about whether we would be able to detect geoengineering if it was being done in secret. Teams of researchers have argued over the scientific possibility of 'testing' geoengineering. How would we measure underlying changes in the reflectivity of the Earth given that the Earth's reflectivity varies all the time in unpredictable ways? Robock *et al.* (2010) have argued that testing of geoengineering would be impossible without its full-scale deployment, in part because the signal of a geoengineering response would get lost in the noise of a chaotic climate system. MacMynowski *et al.* (2011) have responded that we could test the impact of SRM with careful engineering design involving the pulsed switching on or off of geoengineering or the gradual ramping up of deployment. One immediate social scientific contribution to this debate would be to question what is meant by 'test' here. Even if the intended consequences of SRM were immediately detectable, the technology would remain irrevocably experimental. And as one of the SPICE engineers told me, '*You'll never know . . . what the side effects are. We won't even know whether they are side effects [or intended effects].*' The engineer Stephen Salter, who has been involved with geoengineering research for longer than most, claims that 'noise is only a signal which you have not learned to decode yet' (quoted in Hamilton 2013, p. 108). Any climate scientist would respond that some noise was an inevitable and irreducible part of the climate system. Regardless of whether such hubris is realistic, it seems likely that the effect of any geoengineering effort, even at full scale, would not be detectable at least for a matter of years (see Seidel *et al.* 2014). If the technology failed catastrophically, its effects might be felt sooner, but in any case we can anticipate huge disagreements about detection and attribution.

Debates about control among the more vocal geoengineering scientists, and their tendency to drift into using terms such as 'trade-offs' and 'optimisation' (MacMartin *et al.* 2013; Ban-Weiss and Caldeira 2010), worry some of the more ambivalent geoengineering researchers. One of the SPICE team expressed her concerns:

> *'You're going to have to put a load of stuff up there before you see any result and then it's too late to twiddle with it. That's the frightening thing, yes, people will push this, "We can inject here and if we find there are adverse effects then we can just do a little bit of something here and counter that and"*, [Laughs] *It's just crazy to think you could do that.'*

Some more critical geoengineering researchers seek solace in the impossibility of prediction, presuming that the flaws of climate models make responsible deployment an impossibility. Others have taken a different approach to the same realisation, asking what it would take to construct a control mechanism that accounts for uncertainty. A few scientists have begun to ask how we might approach the experimentality of geoengineering, recognising that the substrate would not be multiple possible worlds *in silico* but one very real planet. The control mechanism required for geoengineering would be hugely complicated, but there is the interesting and troubling conclusion from one study that 'the use of SRM need not require a good model of the climate if feedback is used to manage the amount of solar reduction' (MacMartin *et al.* 2014b, p. 256).[2] According to one scientist, the use of SRM in this way would still constitute '*the worst experiment ever*' because of the lack of control and problems of detection and attribution described above. But such research is perhaps a more honest engagement with uncertainty.

Engineering algorithms and control devices force new discussions about what might be considered important and legitimate geoengineering experimentation. But they still divorce the planetary system from any considerations of politics. As I argued in the first chapter of this book, geoengineering is a form of governance, and it is inextricably political. We might therefore consider how people can be put back into the experiments. One recent study, which the researchers were careful to label a 'simulation' rather than an 'experiment', gave a simple climate model and a simple control mechanism to a committee of scientists, asking them to take collective decisions about the appropriate level of geoengineering to restore the model's Arctic sea ice. The 50 years that passed in the model were accelerated to a matter of days, with researchers observing what was happening and taking decisions, at intervals of a 'year' in the model, on how much geoengineering to use (Jackson *et al.* 2013). At the time of writing, the results of this experiment have not been finalised, but a key lesson has already emerged – that doubts, uncertainties and disagreements about both the observed world and desired interventions will be an ever-present feature of geoengineering decisions, even (or perhaps especially) among experts.

Geoengineering, climate models and public credibility

Geoengineering research using models is at an early stage. Scientists' current concerns relate to the ability of models to say useful things about the geoengineering proposals already on the table. There is as yet no consideration of the credibility of these models in public and no consideration of whether, for example, the results of the proposed 'tests' of geoengineering described above might be persuasive outside the scientific community. We can point to the challenges of climate models in the public domain to imagine some of the challenges over public credibility that models might encounter in any future geoengineering scenario. And we can look to the odd but persistent fringe interest in so-called chemtrails (Cairns 2014) – the theory that aeroplane contrails are already being used to, among other things, control the global weather and the minds of individuals – to anticipate how hard it may be to persuade others that something is or isn't being done to the weather.

A key part of the social life of climate models is that they are constructed with particular uses in mind. The imagined uses of climate models – predicting short-term weather or projecting long-term, low-resolution, averaged future climates – allow scientists to tame their endemic uncertainties. For most people, the models are good enough for the job they are being asked to do. With the turn to geoengineering, however, the questions being asked are very different, and a new set of uncertainties is being opened up. At the moment, these uncertainties are largely private. We know from environmental (Sarewitz 2004) and health (Stilgoe 2007) controversies that the public reconstruction of uncertainty can quickly take control of relevant research questions away from scientists.

Geoengineering researchers have started to engage with these endemic uncertainties in new ways, wondering what it would take to geoengineer the world with only a fuzzy picture of the feedbacks between human interventions and a planetary response. To anticipate how models might turn from description to shaping of the worlds they describe, it is instructive to look at economics and finance, in which models are used for both understanding and control. According to MacKenzie and Millo (2003, p. 108), 'Economics does not describe an existing external "economy," but brings that economy into being: economics performs the economy, creating the phenomena it describes' (also see Callon 1998) through the models it constructs. According to MacKenzie and Millo, the familiar critique of economic modelling – that it is based on unrealistic assumptions – misses the point. MacKenzie (2006) takes on Milton Friedman's view that models are 'engines' with which to study the world, rather than 'cameras', but extends it to explain how these engines are now creating new realities, rather than studying them. The explosion of financial capital in the 1980s and 1990s was a product of particular models of finance, and the subsequent implosion in the 2007–8 credit crisis was an example of what MacKenzie and Spears (2014) call 'counter-performativity', in which the use and 'gaming' of the models eventually led to the models becoming disastrously inaccurate. With economics, we see the world being organised to fit the model, rather than vice versa. James Scott has written persuasively about the ambitions of high-modern plans and models to control human beings by making society 'legible' (Scott 1998, p. 2). We might anticipate a similar dynamic if we move towards a geoengineered world. Publics and politics will be impossible to fully encode into model, so the risk is that they are constrained to fit whatever models come to dominate.

With economics, controlled, real-world experimentation is hard, if not impossible. But as with geoengineering, researchers can conduct 'vicarious experiments' in economics, asking 'what if . . . ' (Morgan 2003). Mary Morgan (2003) calls these simulations 'extended thought experiments'. But we should not pretend they are divorced from the real world. Such experiments, like the models in which they are conducted, may shape the world about which they speculate. The 'vast machine' (Edwards 2010) of global climate science risks becoming a vast legitimation machine for geoengineering. The models may go from being tools with which to assess a technology to part of the technology itself.

The comparison with economic models also prompts consideration of politics. There is a broad consensus that, with economics, decisions about interest rates, taxes or public spending can never be purely technical. Although it may be in politicians' interests to pretend that such decisions are largely predetermined, we recognise that there are political choices to be made. Climate science has had to get used to politicisation in organised ways, such as with the IPCC 'Summary for Policymakers', and disorganised ways, as with the 2009 'Climategate' controversy (see Wynne 2010). It is telling that the public deconstruction of climate models is so intense, even with the relatively straightforward conclusion that anthropogenic climate change is real and serious. One can imagine that with models for geoengineering, the complexity of the communication between science and policy will magnify this challenge of credibility impossibly.

The reframing of geoengineering from a phenomenon that can be studied and predicted to an experiment with irreducible uncertainties, in which climate models are not just tools for assessment but part of the governance system, introduces profound questions of experimental ethics and responsibility. Such questions spur the analysis in the next chapter.

Notes

1 See, for example, Vidal (2011).

2 Ken Caldeira has also used control algorithms in model research, which he claims were inspired by experiments in temperature control on an espresso machine (email posted to the Geoengineering Google Group, 31 July 2012).

References

Bala, G., and Caldeira, K. (2000). Geoengineering Earth's radiation balance to mitigate CO_2-induced climate change. *Geophysical Research Letters*, 27(14), 2141–2144.

Ban-Weiss, G. A., and Caldeira, K. (2010). Geoengineering as an optimization problem. *Environmental Research Letters*, 5(3), 034009.

Box, G. E. P., and Draper, N. R. (1987). *Empirical model-building and response surfaces*. New York: Wiley.

Cairns, R. (2014). *Climates of suspicion: 'chemtrail', conspiracy narratives and the international politics of geoengineering*. CGG Working Paper No. 9. Oxford: Climate Geoengineering Governance, Institute for Science, Innovation and Society.

Callon, M. (ed.) (1998). *The laws of the markets*. Oxford: Blackwell.

Clark, N. (2013). Geoengineering and geologic politics. *Environment and Planning A*, 45(12), 2825–2832.

Crutzen, P. J. (2006). Albedo enhancement by stratospheric sulfur injections: a contribution to resolve a policy dilemma? *Climatic Change*, 77(3/4), 211–220.

Edwards, P. N. (2001). Representing the global atmosphere: computer models, data, and knowledge about climate change. In C. A. Miller and P. N. Edwards (eds), *Changing the atmosphere: expert knowledge and environmental governance* (pp. 31–36). Cambridge, MA: MIT Press.

——— (2010). *A vast machine: computer models, climate data, and the politics of global warming*. Cambridge, MA: MIT Press.

Hamilton, C. (2013). *Geoengineering: governance before research please*. Available online at http://clivehamilton.com/geoengineering-governance-before-research-please/ (accessed 7 January 2014).

Hulme, M. (2014). *Can science fix climate change? A case against climate engineering*. Cambridge: Polity Press.

IPCC (2013). Summary for policymakers. In T. F. Stocker, D. Qin, G-K. Plattner, M. Tignor, S. K. Allen, *et al.* (eds), *Climate change 2013: the physical science basis*. Contribution of Working Group I to the Fifth Assessment Report of the Inter governmental Panel on Climate Change. Cambridge: Cambridge University Press.

Jackson, L. S., Crook, J. A., Forster, P., Jarvis, A., Leedal, D., *et al.* (2013). *Simulation of how a geo-engineering intervention to restore arctic sea ice might work in practice*. Abstract No. A33L-04. Presented at the Fall Meeting, American Geophysical Union, San Francisco, 9–13 December 2013.

Kravitz, B., Robock, A., Boucher, O., Schmidt, H., Taylor, K. E., *et al.* (2011). The Geoengineering Model Intercomparison Project (GeoMIP). *Atmospheric Science Letters*, 12(2), 162–167.

Lakoff, A. (2008). The right patients for the drug: pharmaceutical circuits and the codification of illness. In E. J. Hackett, O. Amsterdamska, M. E. Lynch and J. Wajcman (eds), *The handbook of science and technology studies*, 3rd edn (pp. 741–760). Cambridge, MA: MIT Press.

Lenhard, J. (2007). Computer simulation: the cooperation between experimenting and modeling. *Philosophy of Science*, 74(2), 176–194.

Levitt, S. D., and Dubner, S. J. (2010). *Superfreakonomics: global cooling, patriotic prostitutes and why suicide bombers should buy life insurance*. London: Penguin UK.

MacKenzie, D. (2006). *An engine, not a camera: how financial models shape markets*. Cambridge, MA: MIT Press.

MacKenzie, D., and Millo, Y. (2003). Constructing a market, performing theory: the historical sociology of a financial derivatives exchange. *American Journal of Sociology*, 109(1), 107–145.

MacKenzie, D., and Spears, T. (2014). 'A device for being able to book P&L': the organizational embedding of the Gaussian copula. *Social Studies of Science*, 44(3), 418–440.

MacMartin, D. G., Keith, D. W., Kravitz, B., and Caldeira, K. (2013). Management of trade-offs in geoengineering through optimal choice of non-uniform radiative forcing. *Nature Climate Change*, 3(4), 365–368.

MacMartin, D. G., Kravitz, B., and Keith, D. W. (2014a). Geoengineering: the world's largest control problem. In *Proceedings of the American Control Conference (ACC), Portland, OR, 4–6 June 2014* (pp. 2401–2406). Piscataway, NJ: IEEE Service Center.

MacMartin, D. G., Kravitz, B., Keith, D. W., and Jarvis, A. (2014b). Dynamics of the coupled human–climate system resulting from closed-loop control of solar geoengineering. *Climate Dynamics*, 43(1/2), 243–258.

MacMynowski, D. G., Keith, D. W., Caldeira, K., and Shin, H. J. (2011). Can we test geoengineering? *Energy & Environmental Science*, 4(12), 5044–5052.

Morgan, M. (2003). Experiments without material intervention: model experiments, virtual experiments, and virtually experiments. In H. Radder (ed.), *The philosophy of scientific experimentation* (pp. 216–235). Pittsburgh: University of Pittsburgh Press.

Nerlich, B., and Jaspal, R. (2012). Metaphors we die by? Geoengineering, metaphors, and the argument from catastrophe. *Metaphor and Symbol*, 27(2), 131–147.

Orléan, A. (2010). *The impossible evaluation of risk*. Prisme n° 18. Paris: Cournot Centre for Economic Studies.

Pickering, A. (2004). The science of the unknowable: Stafford Beer's cybernetic informatics. *Kybernetes*, 33(3/4), 499–521.

Robock, A. (2012). Will geoengineering with solar radiation management ever be used? *Ethics, Policy & Environment*, 15(2), 202–205.

Robock, A., Bunzl, M., Kravitz, B., and Stenchikov, G. L. (2010). A test for geoengineering? *Science*, 327(5965), 530–531.

Rodhe, H., Charlson, R. J., and Anderson, T. L. (2000). Avoiding circular logic in climate modeling. *Climatic Change*, 44(4), 419–422.

Sarewitz, D. (2004). How science makes environmental controversies worse. *Environmental Science & Policy*, 7(5), 385–403.

Scott, J. C. (1998). *Seeing like a state: how certain schemes to improve the human condition have failed*. New Haven, CT: Yale University Press.

Seidel, D. J., Feingold, G., Jacobson, A. R., and Loeb, N. (2014). Detection limits of albedo changes induced by climate engineering. *Nature Climate Change*, 4(2), 93–98.

Steele, K., and Werndl, C. (2013). Climate models, calibration, and confirmation. *British Journal for the Philosophy of Science*, 64(3), 609–635.

Stilgoe, J. (2007). The (co-)production of public uncertainty: UK scientific advice on mobile phone health risks. *Public Understanding of Science*, 16(1), 45–61.

Sundberg, M. (2009). The everyday world of simulation modeling: the development of parameterizations in meteorology. *Science, Technology, & Human Values*, 34(2), 162–181.

Turney, J. (2013). A model world. *Aeon Magazine*, 16 December.

Vidal, J. (2011). IPCC asks scientists to assess geo-engineering climate solutions. *The Guardian*, 15 June. Available online at http://www.theguardian.com/environment/2011/jun/15/ipcc-geo-engineering-climate (accessed 24 April 2014).

Wynne, B. (2010). Strange weather, again climate science as political art. *Theory, Culture & Society*, 27(2/3), 289–305.

7 The reluctant geoengineers[1]

Coming out of Bristol station, on the way to the university where the Stratospheric Particle Injection for Climate Engineering (SPICE) project is based, you see a pub across the road. It is squeezed into the ground floor of an unlovely modernist office block, but its name and iconography appeal to a proud Bristolian tradition. The pub is the Reckless Engineer, and its swinging sign carries a portrait of Isambard Kingdom Brunel. Brunel's legacy is visible throughout Bristol and in the city's connections with the rest of the world: the Great Western Railway, the Clifton suspension bridge and the SS *Great Britain*, the largest ship of its day, which still rests in Bristol Harbour.

Biographies of Brunel suggest that he would have regarded the nickname with some measure of pride. Brunel would not want to have been regarded as dangerous or negligent, although the hazards of nineteenth century engineering would be unacceptable by today's standards. He yearned for precision in his calculations, experiments and manufacturing. But his combination of scientific expertise, ingenuity, business acumen and political skill personified the Industrial Revolution – looking to change rules rather than just operating within them. One contemporary writer, horrified by Brunel's Great Western Railway, described the engineer as 'an inexperienced theorist, enamoured of novelty, prone to seek for difficulties rather than to evade them, and utterly indifferent as to the outlay which his recklessness entailed upon his employers' (John Latimer, quoted in Buchanan 2006).

Such recklessness, which might now be referred to as 'pushing the limits' (Petroski 2004), would assure his status as a British hero (beaten only by Winston Churchill in a 2002 television poll).[2] Brunel was the engineer who above all reminds us that engineering is not merely applied science, but rather a distinct way of understanding and engaging with the world (see Vincenti 1990). In addition to their different ways of knowing, engineers also bring different ideas of responsibility. In this chapter, I discuss what it means to know about geoengineering from different disciplinary and interdisciplinary perspectives and how this affects the responsibilities of researchers.

Physicists, fantasists and philosophers

An interviewee told me that geoengineering is a construction of '*physicists, fantasists and social scientists*'. In the interests of alliteration, as well as the argument that I made in Chapters 1 and 2, I would suggest changing this to 'physicists, fantasists and philosophers', but the point remains that geoengineering, despite its name, has had relatively little to do with engineering. Just as synthetic biologists have argued that genetic engineering barely counts as engineering (Schyfter 2013), so a real geo*engineer* would look at current geoengineering research with astonishment.

Ralph Cicerone, the president of the US National Academy of Sciences, has argued that, for geoengineering, 'research be considered separately from implementation' and that 'we should proceed as we would for any other scientific problem, at least for theoretical and modeling studies' (Cicerone 2006, p. 223). It is easy to understand scientists' reluctance to engage with questions of technology development, but as scientific research builds up around geoengineering it is hard to sustain the argument that this is equivalent to 'any other scientific problem'. Scientists are in the main hugely ambivalent about geoengineering, but the instinctive response is often to turn their ambivalence into research questions. They will often justify themselves by arguing that they are anti-deployment but pro-research. However, the separation between reluctance and enthusiasm is not so easy. James Fleming is critical of some 'sincere but deluded scientists' who are 'pathologically enthusiastic about their research, but not able to really rein in their enthusiasm' (quoted in Tkacik 2011). I would suggest that there is something more complicated than delusion at work, but I agree that there are dangers in well-intentioned enthusiasm.

If we are to take seriously the idea of geoengineering, we should seek to understand the different methods, cultures, norms, ethics and senses of responsibility across the various disciplines involved. The SPICE project put engineering at the heart of geoengineering research. Before SPICE, knowledge about geoengineering was overwhelmingly *episteme* (knowing that) rather than *techne* (knowing how) (see Ryle 1971; Hansson, forthcoming). The disruption brought by the proposed SPICE experiment can in part be explained by the way in which experiments blend *episteme* and *techne* (Hansson 2014). SPICE has challenged the norms of geoengineering research and exposed underlying assumptions about relevant knowledge and responsibility.

It is hard to put one's finger on the technical novelty of emerging technologies, such as nanotechnology or synthetic biology. Scientists in competing fields may argue that these are merely new labels for established science, as may the nanotechnologists and synthetic biologists themselves when regulators begin asking difficult questions (Rayner 2004). But emerging technologies are often sociologically novel, in the sense of bringing together new combinations of people and knowledges and suggesting new futures.

Interdisciplinarity is one of the characteristics of what Nowotny *et al.* (2001) call mode 2 science, conducted 'in the context of application'. When innovation or policy-relevant knowledge is demanded, innovators or policymakers should not be expected to care much about which disciplines are providing them. With the growing policy emphasis of the accountability and 'impact' of research, disciplines are often talked about as though they are a problem.

We repeatedly hear about the need to break down disciplinary boundaries and break out of disciplinary silos. But the policy rhetoric overlooks the practical difficulties of making interdisciplinarity work – the modes and logics of interdisciplinarity (Barry *et al.* 2008). Interdisciplinarity is not just the mixing of disciplines. The process of integration can see disciplines integrating symmetrically, but there is more likely to be antagonism and subordination as assumptions are shared and negotiated (Barry *et al.* 2008).

When we include matters of experimentation and responsibility, disciplinary assumptions and conventions are vital, particularly in areas such as geoengineering, where the configurations of disciplines are so new. Disciplines are, to use Barry *et al.*'s phrase, 'repositories of a responsible kind of epistemological reflexivity' (Barry *et al.* 2008, p. 26). The responsibilities that geoengineering researchers take on are largely a product of the disciplines that they come from and the conversations and reflections that are produced when disciplines interact.

Searching for the 'perfect particle'

Before the SPICE project began, there was relatively little questioning of what the particles involved in stratospheric particle injection might be. The volcanic analogy meant that the model particles were sulphates. These are produced in large quantities by volcanic eruptions, and they appear to be reasonably good at reflecting sunlight. But the evidence from eruptions is that sulphates also deplete atmospheric ozone (Rasch *et al.* 2008). One strand of the SPICE project is asking whether, once unconstrained by the contents of a volcano, we might be able to find a particle that reflects sunlight better and damages the atmosphere less while remaining airborne long enough to do its job.

The team of atmospheric chemists, climate modellers and volcanologists are looking at, among other materials, titanium dioxide, alumina, silica, calcium carbonate, sea salt, zinc oxide, silicon carbide and diamond dust (Pope *et al.* 2012). They are trying to balance multiple criteria. Small particles are good at dispersing and reflecting, but they are likely to be more reactive. Researchers want something that doesn't condense, doesn't form into clumps and doesn't absorb too much radiation, which would heat up the stratosphere. On the assumption that '*what goes up must come down*', they are also looking at the toxicological and ecotoxicological effects of these chemicals.

Turning this challenge into doable experiments employs the ordinary apparatus of classroom chemistry, as well as cutting-edge 'laser tweezers' to levitate single particles so that researchers can study what might happen to them in the environment. The experimental work required to do these things comes up against problems and serendipities that are a world away from the politics of geoengineering. Measuring the properties of particles requires extensive groundwork just to get apparatus built and working reliably. In one study, a flow tube misbehaves, delaying the research while a new polonium source is found. In another, the laser tweezer experiment has succeeded in trapping a particle for 30 minutes, which they think is a world record. Listening to the individual SPICE scientists present their work, except for the introductions and caveats that bookend their talks, one could be forgiven for thinking their research was taking place in an arcane sub-discipline rather than being a highly politicised issue.

The researchers talk about 'positive' and 'negative' results for their particles, but the work is inching towards the conclusion that there is unlikely to be a 'perfect' particle. Each has its advantages and disadvantages, and once the conversation draws back to include the other SPICE disciplines, the question becomes 'perfect for what?' The engineers begin by expressing their distaste for sulphates. They are thinking about their pipe, and they refer to both sulphuric acid and hydrogen sulphide as '*nasty chemicals*'. They speculate about what might happen if a pipe bursts and 20 tonnes of the stuff falls to the ground. They worry about sulphur dioxide freezing in the pipe and consider whether liquid nitrogen might be a useful carrier for the particle of choice. A discussion about diamond dust raises some eyebrows, but the engineers are surprised to find it is cheaper than they had assumed. It is not the only surprising proposal. David Keith has suggested the possibility of engineering bespoke particles that levitate photophoretically (using solar radiation to stay aloft) (Keith 2010).

The coming together of the disciplines in SPICE challenges the presumption, apparent in much geoengineering research, that there is an optimal solution. There are only better or worse solutions, depending on the criteria considered important at the time. The more disciplines that are involved, the more these criteria proliferate and clash. In particular, the involvement of engineers in SPICE has forced reconsideration of some aspects of geoengineering that had been neglected.

Bounding science and technology

Philosophers of technology have attempted to demarcate science from technology and scientists from engineers according to their goals, success criteria and ethics (Cordero 1998). Walter Vincenti argued that 'for engineers, in contrast to scientists, knowledge is not an end in itself or the central objective of their profession. Rather it is . . . a means to a utilitarian end' (Vincenti 1990, p. 6). Computer scientist and engineer Fred Brooks put it like this: 'A scientist builds in order to learn, an engineer learns in order to build' (quoted in Nightingale 2004). But Vincenti also makes the point that engineering is not just the application of knowledge. Engineers go beyond applied science to generate new sorts of knowledge. According to Edward Constant, 'technology explores the environment directly, not "vicariously" as does science' (Constant 1984). These insights go some way towards explaining what Rabinow and Bennett, in their exploration of synthetic biology, call the 'engineering disposition . . . understanding through making and remaking' (Rabinow and Bennett 2012, p. 15).

An important insight from recent science and technology studies is that the closer one looks at science and technology in areas of innovation, the more they appear to blur into something like 'technoscience' (Latour 1987). The separations between science, engineering and technology may have more to do with the 'boundary work' of the researchers themselves than any essential differences. We can look back to John Tyndall's dismissal of engineering as nothing more than 'trial-and-error' (Gieryn 1983), as well as to current debates around emerging technologies.

Researchers in synthetic biology are explicit about their aim of turning biology into an engineering discipline (e.g. Endy 2005). The dynamics of geoengineering are rather different. There is a caution about embracing engineering and the shifts in responsibility that this would entail. With carbon dioxide removal, engineers have already been involved in doing some of the calculations that have been important in destabilising some of the more hubristic claims for the contribution of direct air capture. Engineers have pointed out that the removal of additional carbon dioxide from the atmosphere with machines would require an industry at least as big as that which put the carbon dioxide there in the first place (e.g. Axon and Lubansky 2012). The contribution of engineering in this case has been to add a technical understanding of the scale of sociotechnical system that might be required to reverse the Industrial Revolution, at speed. One estimate puts the cost of doing this at $1,000 per tonne of carbon dioxide (House *et al.* 2011), meaning that effective removal on a scale to reverse climate change would run to many trillions of dollars.

An early contribution of the SPICE project was to run the numbers on solar radiation management (SRM) options in a similar way. One of the SPICE engineers justified the approach like this:

> *'As an engineer, I don't like anybody talking about something like geoengineering as a . . . potential fix, solution, or whatever, if it turns out that it isn't practical . . . There's no point in going too far down your enthusiasm for a particular course of action if it turns out to be practically impossible.'*

The engineers' paper offered a 'rational comparison of the technologies for delivery of aerosols into the stratosphere' (Davidson *et al.* 2012). It estimated the costs of various options for getting particles into the stratosphere. Certain things are taken into account, such as the costs of new equipment; other things are consciously left out, such as the transport and personnel costs of operating the eventual infrastructure; and other things are ignored, such as the security and terrorism risks of these new, highly centralised, critical facilities, the costs of understanding and compensating for the side effects of deployment, or the possible contingencies and unknown unknowns.

The paper runs down the list of various options that have at some point been proposed for getting substances into the stratosphere: free-flying balloons (disposable or reusable), towers (freestanding or supported by guy ropes), aircraft, artillery firing shells upwards, missiles (disposable or reusable) and airships. These options are evaluated according to their environmental impact and 'social impact'. They are all dismissed, ultimately, as too expensive. The idea of scattering particles using aeroplanes is calculated to be more complicated than first thought. The 'several aircraft' imagined by Budyko (1974) and taken on by Keith

(2013) would, according to the SPICE paper, more likely be a fleet of 1000 planes each flying 1000 trips per year, costing more than £100 billion.[3]

The engineers' serious examination of geoengineering proposals leaves them plenty of room for ridicule. Particular scorn is reserved for the idea of 20-kilometre-high towers. A diagram sketches such a tower, dwarfing not only built structures from the Great Pyramid at Giza to the Burj Khalifa skyscraper in Dubai but also Mount Everest. Here the 'rational assessment' only just stops short of farce. A stratospheric tower made of carbon fibre would have a diameter of more than 500 metres, a weight of 62 million tonnes and a cost of around £250 billion. Given that the current production of carbon fibre is 50,000 tonnes per year, the authors estimate that 'a few trillion pounds would be needed to increase carbon fibre production to the 10+ million tonnes p.a. needed to make the materials for these towers' (Davidson *et al.* 2012, p. 4277).[4] In presenting this work, the engineers joke that such a tower may become cost effective if it were built on top of Mount Everest.

The SPICE paper's final option – 'balloon-supported high-pressure pipes' – comes out on top. A paper published just a couple of years earlier had lumped balloons and towers together as 'more exotic techniques' (Robock *et al.* 2009), but the balloon idea had subsequently been given approval by the authors of *Freakonomics* (Levitt and Dubner 2005), inspired by Nathan Myhrvold's 'Stratoshield' (Intellectual Ventures 2009). The paper calculates the cost of the balloons in the order of billions of pounds. Other options range from tens of billions to trillions.

The engineers working on geoengineering are currently taking on a problem defined at an absurd scale of abstraction. The numbers they use as the starting point for their calculations are the total quantity of carbon dioxide in the atmosphere and the predicted temperature rise. This is climate change-by-numbers, a giant problem that needs to be reduced to a problem conceivable in engineering terms.

If a stratospheric balloon is taken seriously as an engineering proposition, a host of questions demand consideration, and these involve setting aside the political baggage of the issue and reducing the social experiment to a set of technical calculations. Some can be answered in sketches on scraps of paper, as they were to me in conversations with the SPICE engineering team. Others suggest the need for various forms of modelling and testing. If a balloon is to float in the stratosphere, how strong would its tether need to be? The tether would, in effect, be holding down the balloon, but it would need to hold up its own weight, which could be hundreds of tonnes. The balloon and its tether would need to cope not just with the most severe winds and the storms that the troposphere can whip up, but also with the 160 kilometre per hour jet stream that would pummel the middle of the tether once the balloon was at an altitude of 20 kilometres. Would these winds push the tether sideways, meaning 30 kilometres of hose would be needed rather than 20 kilometres? Would they cause vibrations that could wobble the thing to pieces like the Tacoma Bridge? If the tether were expected to carry a liquid inside its whole length, how strong would it need to be? And how could the liquid be stopped from freezing on its journey upwards? All of this would require a pump of extraordinary power – how much would depend on what substance is being pumped up and how much is required.

For the engineers, the various problems and contingencies associated with both climate change and current geoengineering proposals are narrowed such that the balloon itself becomes the problem. The challenge of combatting climate change morphs into the question of how to get a working tethered balloon system. The prospect of a balloon larger than a football stadium, tethered to float above the altitude of an aeroplane, which many imaginations would find ridiculous, becomes to the engineers a fascinating set of research questions.

When the SPICE team assembles, the engineers demand workable information from the climate scientists about high-altitude weather. They want to know more about the jet stream. They ask the climate modellers where gusts will happen, in case they coincide with the natural frequency of the balloon, which could become a giant, upside-down pendulum. They ask the chemists about the health hazards of the various particles under consideration. Each engineering question, once answered, prompts others. If the tether is too heavy, how would it be supported? With more balloons at intervals along the pipe? How big would these need to be? If the balloon needs to be bigger than first imagined, what materials would be needed? How hot would a slurry carrying the particles have to be to stop it freezing?

The engineers' research also throws up tangential questions: How to photograph and detect a long, thin tether as it stretches up into the sky? How to fix their laser vibrometer? They realise the limits of current models, which typically represent a balloon like a soap bubble, with uniform surface tension or, when tethered, as a solid object on the end of a string. When dealing with a balloon of the size they are imagining, these models are deemed inadequate.

The engineers could be seen to be testing this idea to destruction. They would claim that their approach is inherently precautionary; in looking for something that works they have to explore all the reasons why it might not work. They are attuned to some of the limits and social dimensions of technology. But we can see the potential for a form of 'escalating commitment' (Staw 1976) towards the balloon option. Indeed, one of the criticisms of the testbed experiment described in Chapter 5 was that if something went wrong, this would only increase the desire of the engineers to tweak and improve the system.

In interdisciplinary work, different disciplines take some things for granted while exploring others. The climate scientists are investigating the impacts of an imagined technology while taking the technology for granted. The engineers take the problem of climate change for granted, at a level of abstraction that suits their technological interests, while asking what it would take to construct and maintain a technology capable of making a difference. Their reductionism, and the knowledge that it produces, are surrounded by contingencies that are markedly different from those of their natural sciences colleagues.

Intersecting contingencies

In their landmark study of laboratory science, Bruno Latour and Steve Woolgar (1979) trace the production of robust scientific facts from experiments that seem, up close, to be chaotic and contingent. The farther one moves away from the research coalface, the more likely these contingencies are to get erased, a point subsequently developed by Collins (1987) and MacKenzie (1990) in the contexts of public science and technology assessment, respectively. In scientific research, distance seems to lend enchantment.

I described in the previous chapter some of the contingencies of climate models as they apply to geoengineering and how these can get lost as the models are pressed into public service. Climate modellers are normally happy to admit the uncertainties inherent in their code and its depiction of the world, but not everyone would see what Hastrup (2012, p. 6) calls the 'built-in humility' of the models. The climate models may travel independently of the modellers who know their foibles, as one of the SPICE climate scientists described:

> *'The model may get taken and given to a PhD student and they don't really know about it. They kind of run a study, their PhD is reliant on it. You know, there can be a blurring of how much someone knows if it's tuned or not . . . I think there's a lot of overconfidence in quite a few of the modellings that I've seen.'*

Climate models are themselves products of decades of interdisciplinary interactions (Shackley and Wynne 1995), although much of the criticism of current integrated assessment models can be explained by a concern that the various contingencies and concerns of the constituent disciplines, particularly in the social sciences, have been chopped off as they have been assimilated. And, as Steve Rayner describes it, the uncertainties can 'melt into the background' (Rayner 2000, p. 280) when integrated assessment models move away from their origins. An important question for the future of geoengineering research is how the contingencies of various disciplines are maintained and interpreted.

As disciplines are brought together by new issues we see a meeting of their various contingencies, ifs, buts and uncertainties. At first sight, engineers may regard the environmental science as either straightforward, in the sense that we *know* about climate change or *know* about the effect of volcanoes on climate, or impossibly naïve, for example in climate models' inability to predict the real impact of geoengineering. For their part, the natural scientists may begin by regarding engineering questions as either dangerous or irrelevant.

The climate scientists working on geoengineering look at the eruption of Mount Pinatubo and express astonishment that a relatively small amount of material has such a dramatic effect on global climate. The engineers focus not on the small amounts of dust but on what one described as the '*huge amount of energy*' required to propel it into the stratosphere. These two disciplinary views of the same phenomenon lead to very different research questions.

Geoengineering researchers from the natural sciences can criticise what they see as a tendency among engineers to instrumentalise their science, to regard it as a service to a wider project that is, ultimately, an engineering one. When the SPICE team gets together, it is typically the engineers who ask the first questions, who probe the scientists to tell them more about particular aspects of their research. The impression matches that given by Stephen Salter, who told Jeff Goodell that 'scientists are people who know more and more about less and less, while engineers have to know a little bit about a lot of things, and they have to learn it fast' (quoted in Goodell 2010, p. 167). The scientists' suspicion is that the engineers are working in the wrong direction, starting from the device and selecting whichever pockets of understanding help them.

For some scientists, the differences in approach are less about epistemology than ethics:

> *'Engineers generally want to do stuff. Scientists are content to understand it without necessarily wanting to do anything . . . Engineers tend to be more pragmatic and less constrained by broader considerations. They tend to regard the question of whether you can do it as entirely separate from the question of whether you should do it . . . Scientists would tend to regard knowledge as value free and engineers would tend to regard technology as value free.'*

One of the SPICE team saw this as a difference in disposition: '*They think everything's possible and people can fix things, and I think the scientists tend to be more doubtful of our abilities to solve anything in the real world and more doom-and-gloomy.*'

These differences are perhaps most clearly expressed in the ways in which geoengineering researchers express and regard their own responsibilities.

Responsibility and uncertainty

A geoengineered world would in principle mean humanity taking responsibility for the climate, if not for every weather event. It would mean a massive extension of the modernist project of joining risk to blame. Every storm, flood or drought could be redrawn as somebody's fault. 'Acts of God' would become points of litigation. However, following the arguments in my first two chapters, we should not leapfrog more immediate discussions of responsibility. In engaging with the uncertainties of the geoengineering debate, researchers also engage with a set of new responsibilities.

Engineering, as a profession and a research discipline, defines itself in relation to, rather than apart from, the society that it purports to serve (Nichols and Weldon 1997). Engineering, like medical science, attends to human needs, even if these needs are not straightforwardly defined. With the growth of engineering as a profession during the twentieth century, engineering ethics began to be taught in universities, in much the same way as medical ethics became a part of doctors' training. Questions of responsibility are unavoidable in engineering, although they tend to be answered in rather a narrow way. Much of engineering ethics could be characterised as 'disaster ethics' (Kline 2001), aiming to avoid catastrophe or to reassure the public, rather than reflecting on bigger questions of the common good.

At one SPICE meeting, a small section of aerodynamic hosepipe was passed around the group. The model was a low-tech mock-up, but it was still a designed *thing*, inserted into a debate that remains resolutely immaterial. Its appearance at the SPICE meeting was met with confusion. The natural scientists were unsure how to engage with this particular toy.

Even when they are not making things, the engineers are seeking to understand a future world of things made for the purpose of geoengineering. To some of their non-engineer colleagues, this feels premature. The engineers on the SPICE project can, when challenged, retreat to a well-established response in which their role is not to question problems, but to provide solutions. They separate questions about what is possible from those about what is desirable. SPICE researchers told me about the need to consider practicalities before ethics and about engineers' foremost responsibilities towards their 'client'.

The Geoengineering Google Group – an email forum – bears witness to some examples of engineers going much further, expressing contempt for ethics as they imagined it. In an email directed at the philosopher Ben Hale, one engineer responds to Hale's suggestion that experimentation with geoengineering might pose ethical concerns:

> You are totally wrong and confused. There is a vast range of philosophical questions associated with implementation of geoengineering but the issue is not implementation. Who are you to tell geoengineers what they can think about, what they can calculate, when they can do small, controlled, safe 'laboratory' experiments in which peers have an input. I suppose you would like to have such control but you won't get it.
>
> (Email, 12 November 2012)[5]

In a subsequent message:

> Geoengineers develop options. If they move into implementation then they assume a different role where ethics can play a part. However, ethics has nothing to say about Geoengineering R&D other than 'do no harm'.
>
> (Email, 12 November 2012)

And, following a suggestion that geoengineering researchers themselves need a better understanding of philosophy:

> What an outrage! Should engineers apply philosophical theory to determine what they study? In any case politicians and related will decide what is implemented, not the engineers. Most politicians are unimpressed by ethics. Tell the ethicists to stuff it; but in any case leave geoengineering alone and focus on the decision makers.
>
> (Email, 12 November 2012)

It would be wrong to read such views as in any way representative of the geoengineering research community. The Geoengineering Google Group was initially overseen by Ken Caldeira, but this role has now moved to an enthusiastic amateur. The forum is open to non-scientists, and its archives are visible to anyone who cares to search, but some of its contributors write as though they are unaware of or uninterested in its transparency. It has an information-sharing function, announcing relevant new papers or opinion pieces. It also provides a forum, particularly for those who are not regular participants in the grand tour of geoengineering meetings, to make sense of and shape the politics of geoengineering. The most vocal participants are those advocating a particular geoengineering option, such as biochar, or those who see climatic emergencies as demanding urgent geoengineering. The public ambivalence of the mainstream geoengineering research community is only occasionally represented. For this reason, one of the SPICE researchers described the email forum as '*a slow motion car crash*'.

Carl Mitcham (1997) argues that the model in which engineers are responsible only to their clients evolved over the twentieth century to take account first of engineers' own value considerations and subsequently of their wider social responsibilities. Further conversations with the SPICE engineers trace this evolution. One of them admitted that in this case, the client could be imagined only in abstract terms as the human race or the planet. They therefore admitted the need to consider their social responsibilities in addition to their role responsibilities (see Douglas 2003).

Engineering is a design discipline. It is explicitly normative in the way it imposes particular imaginations onto the world.[6] The presumption of particular technological fixes first demands the articulation of a problem. For Petroski, 'understanding and articulating the problems with the existing system . . . is essential to working out an engineering solution to a problem' (Petroski 1996, p. 147). As one of the SPICE engineers put it, '*There's a problem that needs to be solved so the engineers get to it and try to solve it . . . It's just a standalone problem.*'

According to Downey *et al.* (2006, p. 108), 'drawing a boundary around a problem was the essential step in learning the "engineering method"' in the post-war United States, but they admit that other cultures can and should imbue engineering practice with additional values. The growth of curricula and research within 'engineering ethics' points to a recognition that engineering is about more than 'right or wrong answers' (Downey *et al.* 2006, p. 108).

But the definition, and therefore narrowing, of a problem is a problematic negotiation. Engineers develop things, but these things have imagined functions. Questions of what a technology is and what it is for are entangled (Kroes 2010). Engineers know that technologies can be remixed and repurposed. A stratospheric balloon may turn out to be more useful for global communications than for particle spraying, so the engineers are ready to switch targets and shape another doable problem. But they cannot completely escape the substantive aspects of their technologies, in which their designs embed particular values, through particular definitions of the problem. Technologies are never just means to ends. They are detours, new paths taken towards particular definitions of problems (Latour and Venn 2002).

An important insight from literature in the social construction of technology is that the direction in which technologies evolve depends on how individuals and groups represent the problems to which the technology is a solution. At one level, the problem to which geoengineering offers a solution – climate change – seems uncontroversial. But the problem imagined by the SPICE engineers is more specific – how to get thousands of tonnes of a particular material into the stratosphere. Offering to solve this 'problem' is much more controversial. There are plenty of geoengineering researchers, within and outside SPICE, who think that asking this question is itself irresponsible or a distraction from more important agendas.

Those who see climate change as an engineering problem (see Chapter 3) and the subset who see the task of stratospheric aerosol injection as a more well-defined engineering problem may argue that we therefore require something like an Apollo programme or a Manhattan Project to get the job done. But as David Collingridge described in 1980, it was easy to tell when these projects had worked: 'Success for the Manhattan Project was a bomb which exploded with more than a particular force. Success for the moon programme was the landing of a piece of hardware carrying a man and its safe return to Earth' (Collingridge 1980, p. 15). When it comes to schemes such as the Green Revolution, which sought to improve crop yields around the world, success is more contested, not least because of vast disagreements about what 'the problem' was (Collingridge 1980).

The more time the SPICE researchers spend working on geoengineering, the more the analogies with nuclear bombs and moonshots unravel. One SPICE engineer referred to the task of designing and building a nuclear weapon as '*actually a fairly confined problem . . . a fairly simple problem compared with trying to understand the planet*'. Nevertheless, they believe that engineering should be part of the response.

The natural scientists within SPICE are less sure. Before the project began, some of them viewed geoengineering as a dangerous aberration. One told me, '*I'd thought about it. I'd heard about it. I didn't like it.*' But in common with the engineers, and partly as a result of the unusual way in which the project was funded, most were new to the debate. Some were resolute in sticking to their disciplinary research, with one responding, '*The science, I'm very interested in. It's the politics that goes with it . . . that's not what I'm expert in.*' However, this same scientist, a climate modeller, still justified getting involved in terms of tempering others' use of climate models to justify stratospheric geoengineering:

> '*I suppose I've gone into it wanting to demonstrate that we can't rely on a model to do something like that . . . It needs people like me to be engaged . . . It's very interesting to be forced to think about [geoengineering] because, even being within the [SPICE] project, it's much easier to just say "Ooh, I'm just learning about volcanoes, we're just making the models better", and it's very easy to put blinkers on.*'

Postdoctoral researchers and PhD students who joined the project once it was underway revealed similar concerns:

> '*I was very uncomfortable with taking the job in the first place because it was geoengineering and my feeling about geoengineering was it was nuts and we shouldn't*

> *be messing with the natural ecosystems because we know so little about them . . . I came to realise that people were actually taking these geoengineering ideas seriously, which was even more terrifying to me and that sort of made me split; half of me wanted to run screaming in the other direction and have nothing to do with it but the other half of me felt that it was really important that people take a good hard look at these technologies.'*

Another junior SPICE researcher told me this:

> *'Working on geoengineering is quite exciting compared to normal climate science . . . you kind of feel like you're actively contributing science towards policy decisions . . . I'm one of those scientists that thinks the ethics actually kind of comes first . . . So if you're trying to do good in a study you should think about ethics before you do that study.'*

But this scientist also said, '*It'd be nice not to have geoengineering directly in the study*', arguing that the abnormality of the politics interferes with the normal process of doing science. Younger geoengineering researchers appreciate the career risks of defining themselves in such narrow terms. They see a tension between disciplinary allegiance and the need for interdisciplinarity as they '*chase geoengineering's tail*', as one put it. One of the SPICE scientists echoes a common scientific refrain: '*I just want to figure out what the answer is*'. But the SPICE team recognise that some of the questions are so strange that the act of answering them can never be straightforward.

SPICE researchers appreciate that by getting involved in geoengineering they may be inadvertently and incrementally contributing to the reification of the idea that they find so problematic. Their project is already locked into the particular idea of stratospheric particle injection, although one could argue that this concentration is the result of previous assessments from the Royal Society and others (see Bellamy *et al.* 2013). In news stories and on websites, the iconography of the full-scale stratospheric balloon attached to a giant ship is inescapable. Their concern is that their research will increase the possibility of further lock-in. Matt Watson, the SPICE lead researcher, described on his blog, while preparing the response to the Engineering and Physical Sciences Research Council (ESPRC) stage-gate, coming to terms with this dynamic and the importance of steering clear of intentional or accidental advocacy of single geoengineering options.[7]

The decision to research geoengineering, even if justified as a defensive response to the misplaced technological enthusiasms of others, legitimates a set of proposals that until very recently had been seen as fantasy. One scientist sensed that '*the pressure's on to keep the controversy going*'. But the team's discussions over the course of the project also reveal an emerging sense of responsibility. They are quite open to the possibility that their research could destabilise the object of their attention, making stratospheric particle injection less, not more, likely by demonstrating its difficulties and opening up new uncertainties.

Caveats and the public control of uncertainty

Most of the climate modellers who have turned their attention to geoengineering recognise that even if the scientific tools are the same and the questions only incrementally different from those of conventional climate science, geoengineering research has a radically different public and political context. They are aware of the need to, as one put it, '*get results, communicate them and communicate them carefully, because . . . they are still just models*'. However, in the absence of any other experiments, the claims made from model studies have been hugely important in shaping the geoengineering debate. The scientists who first ran simulations of a geoengineered world admit that they wanted to demonstrate what a bad idea it would be. (Some go on to describe how the results of these model runs surprised them by suggesting it wouldn't be as bad as they feared.) As in the more mature debate on climate models and climate change, accusations of under- and over-claiming from models runs abound.

In geoengineering research, models have been asked to do more than just imagine future climates. Models have been asked to project the implications for various systems in a hypothetical geoengineered world. Over the few years in which I have been interacting with them, geoengineering researchers have displayed a growing confidence in talking about 'winners and losers' from geoengineering. This kind of talk was previously reserved for private scientific discussions, at which discussions of the researchers' modelled worlds become playfully unreal. The researchers talk in private about the dangers of 'switching off the Indian monsoon', 'greening the Sahel', 'warming the poles' or 'wiping out Mali' as though the planet were a toy. In doing so, they typically also imagine, explicitly or implicitly, either a benign global dictator or a homogeneous collective – Mike Hulme's 'global "we"' (Hulme 2014, pp. 54–55) – that will be making the decisions.

Geoengineering researchers discuss the calculus of winners and losers even though they know that with tiny adjustments to parameters the balance might flip in different models, with different conditions. At times, they prompt each other to remember that there is a real world to consider and that they should not pretend to be more certain about geoengineering than they are about climate change. In public, they are cautious. Scientific papers drawing on models are typically grounded with the judicious use of caveats that advise against drawing strong conclusions. These caveats cannot, however, completely control the public interpretation of their claims or manage the public reconstruction of uncertainty.

When Cold War scientists began to consider SRM, they imagined that among other beneficial side effects, crop yields would increase. The thinking was that crops would benefit from the continued increase in atmospheric carbon dioxide, without suffering from increased temperatures. Hamilton (2013, p. 252) quotes Harrison Brown as saying that 'if in some manner the carbon dioxide context of the atmosphere could be increased threefold, food production would be doubled', a claim later endorsed by some on the right of US politics. In response to speculation that geoengineering might threaten food and water security, some scientists have run their models to project crop yields. The conclusion of one study – that global crop yields would increase in a geoengineered world – is given the sort of caveat that has become obligatory in geoengineering modelling studies:

> Therefore, although SRM may allow beneficial effects of CO_2 fertilization at a comparatively low level of climate change, the potential for such approaches to reduce the overall risks is still far from established. The safest option to reduce the climate risks to global food security may be to reduce emissions of greenhouse gases.
>
> (Pongratz *et al.* 2012, p. 101)

The paper also argues that 'to cover the full range of uncertainties, future studies should be carried out that employ a wider range of crop and climate models' (Pongratz *et al.* 2012, p. 104). This statement betrays a confidence in models and their ability to capture 'the full range of uncertainties' that is shared by David Keith. Keith says that for stratospheric particle injection, there is 'very strong evidence that it can substantially reduce many of the most important climatic changes and their associated risks' (Keith 2013, pp. 8–9). He adds a note of caution about scientific uncertainty but goes on to say that 'crop productivity in some of the hottest – and poorest – regions of the world would be higher with an appropriate amount of geoengineering' (Keith 2013, p. 9). Keith concludes that 'the balance of evidence from the climate models used to date suggests that doing a little bit would reduce climate risks' (quoted in O'Donnell 2013).

Another analysis by a mixed science and social science team uses model results to discuss the 'optimisation' of SRM in terms of 'inequalities', 'effectiveness' and 'winners and losers' (Moreno-Cruz *et al.* 2012). The paper ends with this caveat: 'Our results, of course, rely heavily on the GCM [general circulation model] data we use in our analysis and should be observed in the context of the significant uncertainties associated with these models' (Moreno-Cruz *et al.* 2012, p. 662). Another paper, on geoengineering as a problem of optimisation, states, 'However, attempts to intervene in the climate system present a wide range of serious environmental and socio-political risks, a thorough discussion of which is beyond the scope of this study' (Ban-Weiss and Caldeira 2010).

According to David Rier (1999), caveats are used by scientists to manage the risks of interpretation by their peers and the public. But the epidemiologists interviewed by Rier downplayed the public role of caveats. Caveats are more likely to be used by scientists as tools for the management of their credibility within their own community. The geoengineering research community has a porous boundary between public and private, so the norms for the communication of model results and their caveats are not well established. Some scientists are happy to represent model results in public. Ken Caldeira tweeted that 'models suggest people most vulnerable to climate change may benefit most from geoengineering. Risky desperation'.[8] This tweet was in response to a piece in *The New York Times* by Naomi Klein. Caldeira went on to criticise Klein for her 'factual errors', which apparently included her claim, from previous modelling studies, that 'mimicking the effects of a volcano would interfere with monsoons in Asia and Africa, potentially threatening water and food security for billions of people' (Klein 2012).

The history of technology would suggest that advanced technologies, especially centralised sociotechnical systems controlled by rich people, will tend to exacerbate the gap between rich and poor rather than close it. Claims to predict winners and losers might be moderated with the additional uncertainties that would come from dependence upon an inherently unpredictable, highly politicised sociotechnical system. As we move away from the research face, the range of uncertainties that might be considered relevant multiplies (see MacKenzie 1993). We see, through their use of caveats, scientists' attempts to maintain control of these uncertainties.

Following the controversy over the outdoor experiment, the SPICE team have grown increasingly aware of the various audiences to which their research might speak. One of the SPICE team was involved with a paper that discussed how 'asymmetric forcing from stratospheric aerosols impacts Sahelian rainfall' (Haywood *et al.* 2013). The research was framed in terms of the possible effects of volcanic eruptions on droughts in the Sahel. But the experiments that were conducted, using the Met Office's HadGEM (Hadley Centre Global Environment Model) climate model, simulated both volcanic eruptions and stratospheric geoengineering, with the only modelled difference being that volcanoes were assumed to pump up particles once, while geoengineering was assumed to be ongoing. The conclusion was that sulphates put into the stratosphere in the northern hemisphere, whether from geoengineering or from a large volcano, would cause drought in the Sahel. If particles were injected in the southern hemisphere, the Sahel would get more rain.

The paper's findings could have been interpreted any number of ways. The conclusions of the paper could, in the hands of more hubristic scientists, be taken as evidence of the possible benefits of well-placed geoengineering. The paper attracted media attention, not least because the lead researcher was located at the Met Office, an institution that plays multiple important roles: policy adviser and scientific research institute, as well as weather forecaster. But thanks to careful wording, the paper's conclusion that there was a need for global governance of geoengineering deployment became the headline.

It was left to ETC Group to draw out some more-speculative implications. Conflating the Met Office's research and policy functions, ETC Group saw the study as a step towards a world in which geoengineering would be used by Western governments as a form of 'foreign aid', steering regional climates away from drought. Even if the imagined endpoint was fanciful, ETC's analysis of the dynamics of mission creep and normalisation in geoengineering research serves as a useful warning (ETC Group 2013).

Some of the SPICE team were flattered to have their work referred to by Raymond Pierrehumbert at the American Geophysical Union. As others in conventional climate science had done when Paul Crutzen published his paper (Crutzen 2006), Pierrehumbert reminded his audience of the high-level problems with SRM, calling it 'barking mad'. He went on to argue the following:

> We don't actually know enough about aerosol formation and about response of models to aerosols to begin doing this kind of fine tuning to even figure out how much we should put up there. There's some very good work [as part of SPICE] . . . that shows how actually the modelling technology is not even up to doing this adequately despite what some aggressive proponents of geoengineering say.
>
> (Pierrehumbert 2012)

Ken Caldeira's reply, in the form of an email to the Geoengineering Google Group, was to draw the conventional distinction between those who support geoengineering and those who support *research* on geoengineering. He argued that modellers knew the foibles of models and were careful to add caveats to their claims.[9]

Some within the SPICE team are uncomfortable with hedging scientific overstatement by using caveats. Their suspicion is that some within the geoengineering modelling community believe that the models are able to reliably predict the effects of geoengineering and that the use of caveats at the end of papers is merely a way to cover scientists' backs:

> *'Say if you were to do a study and its title was "Controlling the Indian monsoon through use of sulphate aerosols". Let's say that's the title and you've done your study and you've concluded you can control the Indian monsoon really nicely if, whatever . . . Even if the scientists will appreciate those caveats . . . I think it can be quite dangerous if that's just limited to one paragraph at the end . . . I don't think you can say that every study is neutral by putting caveats in at the end.'*

These caveats do not change the framing of the research, which is that the implications of SRM are amenable to empirical analysis, prediction and adjustment, nor does it disguise the palpable excitement that some within the climate change research community feel at the ability to use their tools for new purposes.

An alternative approach would be, as one SPICE modeller put it, to '*shout the uncertainties from the rooftops*' in the way that research is done and communicated. The SPICE modellers see a clear need to speak back to the imperfections of the models as much as to the outside world. Some of the people running geoengineering experiments on climate models, such as Alan Robock, have already declared an a priori opposition to the idea of geoengineering, and yet the pattern has been for these studies to emphasise the knowns and downplay the unknowns. When uncertainty is discussed, it tends to be as justification for further research, leading to improved resolution of the assessments, rather than as something that fundamentally undermines the attempt to predict. One of the SPICE scientists reflected that over the life of the project, he had become more uncertain and critical of those who express sufficient confidence in the models to predict either that geoengineering will in some way 'work' or that it is doomed to fail:

> *'You can parrot these things off. It's the geoengineering researcher's mantra: "There's no magic bullet, only winners and losers". But very few people*

> *actually think deeply, I would argue, about what that actually means . . . I think instinctively and on an evidence basis it's pretty obvious that the people that are going to be worst affected by climate change, I would suggest, would probably be the people that will struggle with [climate change] adaption and with geoengineering.'*

This scientist goes on to talk about two figures who have been prominent in geoengineering discussions, both of whom express huge confidence in their science but end up with very different conclusions:

> *'I think at the start of the project my mindset would have been to find out which one of these two eminent scientists is correct and now what I think is I don't trust either of them because I don't think we know enough. You know, everything they say is caveated and they're very clever but . . . I don't want people who have a strong opinion about things when there's so much uncertainty . . . anybody that's got a strong opinion is on shaky ground . . . The thing that SPICE has done to me is it's made me comfortable with not knowing things.'*

Working together

As I described in Chapter 5, the reaction to the SPICE outdoor experiment suggested a paradoxical concern, shared by some within SPICE, about the involvement of engineers in geoengineering. A closer look reveals that engineers are playing multiple roles. They are interested not just in making things, but in understanding what it would require to exert control, through engineering, over the climate. The relative absence of engineering considerations from geoengineering has contributed to a discourse of inevitability. Little thought has been given to the creation of either the hardware or the operating system software required for climate control. The technology is assumed to be already here, on the horizon or 'just around the corner' (Evans *et al.* 2009). Uncertainties, from this view, are seen as clearly defined and resolvable through scientific research.

The different disciplines gathered within SPICE have all in their own ways tried to carve doable problems (Fujimura 1987) from a still nebulous issue. To make geoengineering thinkable in terms of their discipline, each has approached geoengineering with its own norms, assumptions, contingencies and sense of responsibility. Their research questions approach geoengineering obliquely rather than directly. Research on geoengineering is not just about geoengineering. SPICE is not just about stratospheric particle injection for climate engineering. It is also about improving climate models, advancing the use of laser tweezers in experimental chemistry and understanding the dynamics of tethered balloons.

These disciplinary commitments are resilient, but they have begun to be challenged in the conversations that the SPICE researchers have had among themselves and with the outside world. Asked about the stakeholder engagement exercise that was imposed on the project by EPSRC and the stage-gate panel, one of the SPICE team said, '*I thought it was a necessary evil to start with but I think beyond that now, and I think actually it's rather important.*' The SPICE scientists, in the main, regarded public and stakeholder engagement in the same way as they regarded engagement with each other (and with me). Towards the end of the project, one SPICE scientist reflected on the work required to talk across disciplinary cultures:

> *'We've suddenly really got much, much better at talking to each other and, although the social science is another challenge, actually the challenge of the engineers and the modellers and the physical scientists talking to each other is a profound challenge in its own right.'*

The experiment in interdisciplinarity that is happening within SPICE presents the possibility of a more critical, reflexive geoengineering research. Harvey Brooks argued that a vital role for science is as 'the conscience of technology' (Brooks 1994). As disciplinary contingencies interweave, assumptions are challenged and the dominant framing of geoengineering – it is doable and effective but risky, with calculable winners and losers – is challenged. One of the SPICE team reflected on where SPICE had ended up:

> *'I would suggest that the vast majority of what SPICE will publish will suggest it's not as good an idea as we thought it was . . . the ozone stuff is pretty alarming; titanium's a non starter, it's just so photo-reactive it's just insane to even be thinking about it . . . the pipes are going to be difficult . . . and if you do this badly hemispherically [unbalancing the northern or southern hemispheres] you have profound effects . . . I think when the dust has settled, I think actually SPICE will have made stratospheric geoengineering, particularly by tethered balloons, slightly less likely, certainly in the way it was originally considered.'*

Perhaps the greatest contribution of SPICE will be not in building knowledge about geoengineering, but in helping to destabilise the particular imaginary of stratospheric particle injection, and so break the cycle of speculation that has been allowed to dominate geoengineering discussions.

Notes

1 This chapter's title is stolen from the name that Matt Watson, the SPICE principal investigator, gave to his personal blog: www.thereluctantgeoengineer.blogspot.co.uk.
2 There were allegations at the time that the election had been skewed by the concerted efforts of students from Brunel University in West London.
3 The 1992 report from the US National Academy of Sciences saw Navy guns as being the best way of getting particles into the stratosphere, dismissing aircraft as 'impractical' (NAS 1992, p. 453). In another moment of delicious understatement, the report also dismissed a proposal to use billions of small reflective balloons as 'somewhat unattractive' (p. 454).
4 MacKerron (2014) adds that cost estimates normally ignore the market effects of such a dramatic increase in demand, most importantly the likely price increases.
5 Email posted to the Geoengineering Google Group.

6 Science and technology studies researchers argue that the sociotechnical imaginaries of research scientists do the same thing, albeit less explicitly.
7 See posts at www.thereluctantgeoengineer.blogspot.co.uk: '5 reasons why lock in is unlikely in SRM geoengineering' (5 June 2011); 'Separation of science and policy?' (3 June 2011); 'Advocacy promotes lock in' (5 June 2011); and 'How to encourage lock in' (14 June 2011) (accessed 23 May 2014).
8 Caldeira, K. [kencaldeira], Twitter post, 5 November 2012, 3:07 PM, https://twitter.com/KenCaldeira/status/265621333923205120.
9 Caldeira, K., email posted to the Geoengineering Google Group, 9 December 2012.

References

Axon, C. J., and Lubansky, A. S. (2012). *Costing the Earth: resource requirements for carbon dioxide removal schemes*. Presented at the Planet Under Pressure Conference, London, 26–29 March 2012.

Ban-Weiss, G. A., and Caldeira, K. (2010). Geoengineering as an optimization problem. *Environmental Research Letters*, 5(3), 034009.

Barry, A., Born, G., and Weszkalnys, G. (2008). Logics of interdisciplinarity. *Economy and Society*, 37(1), 20–49.

Bellamy, R., Chilvers, J., Vaughan, N. E., and Lenton, T. M. (2013). 'Opening up' geoengineering appraisal: multi-criteria mapping of options for tackling climate change. *Global Environmental Change*, 23(5), 926–937.

Brooks, H. (1994). The relationship between science and technology. *Research Policy*, 23(5), 477–486.

Buchanan, R. A. (2006). *Brunel: the life and times of Isambard Kingdom Brunel*. London: Hambledon Continuum.

Budyko, M. I. (1974). *Izmeniya klimata*. Leningrad: Gidrometeoizdat. (Translation: 1977, *Climatic changes*, Washington, DC: American Geophysical Union.)

Cicerone, R. J. (2006). Geoengineering: encouraging research and overseeing implementation. *Climatic Change*, 77(3), 221–226.

Collingridge, D. (1980). *The social control of technology*. Milton Keynes, UK: Open University Press.

Collins, H. M. (1987). Certainty and the public understanding of science: science on television. *Social Studies of Science*, 17(4), 689–713.

Constant, E. W., II (1984). Communities and hierarchies: structure in the practice of science and technology. In R. Lauden (ed.), *The nature of technological knowledge: are models of scientific change relevant?* (pp. 27–46). *Sociology of the Sciences Monographs*, Vol. 4. Dordrecht: Springer.

Cordero, A. (1998). On the growing complementarity of science and technology. *Society for Philosophy and Technology*, 4(2), 11.

Crutzen, P. J. (2006). Albedo enhancement by stratospheric sulfur injections: a contribution to resolve a policy dilemma? *Climatic Change*, 77(3/4), 211–220.

Davidson, P., Burgoyne, C., Hunt, H., and Causier, M. (2012). Lifting options for stratospheric aerosol geoengineering: advantages of tethered balloon systems. *Philosophical Transactions of the Royal Society A: Mathematical, Physical & Engineering Sciences*, 370(1974), 4263–4300.

Douglas, H. E. (2003). The moral responsibilities of scientists (tensions between autonomy and responsibility). *American Philosophical Quarterly*, 40(1), 59–68.

Downey, G. L., Lucena, J. C., Moskal, B. M., Parkhurst, R., Bigley, T., *et al.* (2006). The globally competent engineer: working effectively with people who define problems differently. *Journal of Engineering Education*, 95(2), 107–122.

Endy, D. (2005). Foundations for engineering biology. *Nature*, 438(7067), 449–453.

ETC Group. (2013). *Normalizing geoengineering as foreign aid.* Ottawa: ETC Group. Available online at http://www.etcgroup.org/content/normalizing-geoengineering-foreign-aid (accessed 13 September 2013).

Evans, R., Kotchetkova, I., and Langer, S. (2009). Just around the corner: rhetorics of progress and promise in genetic research. *Public Understanding of Science*, 18(1), 43–59.

Fujimura, J. H. (1987). Constructing 'do-able' problems in cancer research: articulating alignment. *Social Studies of Science*, 17(2), 257–293.

Gieryn, T. F. (1983). Boundary-work and the demarcation of science from non-science: strains and interests in professional ideologies of scientists. *American Sociological Review*, 48(6), 781–795.

Goodell, J. (2010). *How to cool the planet: geoengineering and the audacious quest to fix Earth's climate*. Boston: Houghton Mifflin Harcourt.

Hamilton, C. (2013). *Earthmasters: the dawn of the age of climate engineering*. New Haven, CT: Yale University Press.

Hansson, S. O. (2014). *Experiments: why and how?* Presented at the New Technologies as Social Experiments Conference, Delft University of Technology, Delft, Netherlands, 24–25 January 2014.

——— (Forthcoming). Experiments before science? What science learned from technological experiments. In S. O. Hansson (ed.), *The role of technology in science: philosophical perspectives*. Dordrecht: Springer. (In press.)

Hastrup, K. (2012). Anticipating Nature. In K. Hastrup and M. Skrydstrup (eds), *Routledge studies in anthropology*, Vol. 8: *The social life of climate change models: anticipating Nature* (pp. 1–29). London: Routledge.

Haywood, J. M., Jones, A., Bellouin, N., and Stephenson, D. (2013). Asymmetric forcing from stratospheric aerosols impacts Sahelian rainfall. *Nature Climate Change*, 3(7), 660–665.

House, K. Z., Baclig, A. C., Ranjan, M., van Nierop, E. A., Wilcox, J., and Herzog, H. J. (2011). Economic and energetic analysis of capturing CO_2 from ambient air. *Proceedings of the National Academy of Sciences*, 108(51), 20428–20433.

Hulme, M. (2014). *Can science fix climate change? A case against climate engineering*. Cambridge: Polity Press.

Intellectual Ventures (2009). *The stratospheric shield: a practical, low-cost way to reverse catastrophic warming of the Arctic – or the entire planet.* White Paper. Bellevue, WA: Intellectual Ventures Lab. Available online at http://www.intellectualventureslab.com/assets_lab/Stratoshield-white-paper-300dpi.pdf (accessed 1 December 2014).

Keith, D. W. (2010). Photophoretic levitation of engineered aerosols for geoengineering. *Proceedings of the National Academy of Sciences*, 107(38), 16428–16431.

——— (2013). *A case for climate engineering*. Cambridge, MA: MIT Press.

Klein, N. (2012). Geoengineering: testing the waters. *New York Times Sunday Review*, 27 October. Available online at http://www.nytimes.com/2012/10/28/opinion/sunday/geoengineering-testing-the-waters.html (accessed 1 July 2014).

Kline, R. R. (2001). Using history and sociology to teach engineering ethics. *Technology and Society Magazine, IEEE*, 20(4), 13–20.

Kroes, P. (2010). Engineering and the dual nature of technical artefacts. *Cambridge Journal of Economics*, 34(1), 51–62.

Latour, B. (1987). *Science in action: how to follow scientists and engineers through society*. Cambridge, MA: Harvard University Press.

Latour, B., and Venn, C. (2002). Morality and technology: the end of the means. *Theory, Culture & Society*, 19(5/6), 247–260.

Latour, B., and Woolgar, S. (1979). *Laboratory life: the social construction of scientific facts*. London: Sage.

Levitt, S. D., and Dubner, S. J. (2005). *Freakonomics: a rogue economist explores the hidden side of everything*. New York: HarperCollins.

MacKenzie, D. (1990). *Inventing accuracy: a historical sociology of nuclear missile guidance*. Cambridge, MA: MIT Press.

——— (1993). *Inventing accuracy: a historical sociology of nuclear missile guidance*. Cambridge, MA: MIT Press.

MacKerron, G. (2014). *Costs and economics of geoengineering*. CGG Working Paper No. 13. Oxford: Climate Geoengineering Governance, Institute for Science, Innovation and Society. Available online at http://www.geoengineering-governance-research.org/perch/resources/workingpaper13mackerroncostsandeconomicsofgeoengineering.pdf (accessed 18 July 2014).

Mitcham, C. (1997). Engineering design research and social responsibility. In K. Shrader-Frechette and L. Westra (eds), *Technology and values* (pp. 261–278). Lanham, MD: Rowman & Littlefield.

Moreno-Cruz, J. B., Ricke, K. L., and Keith, D. W. (2012). A simple model to account for regional inequalities in the effectiveness of solar radiation management. *Climatic Change*, 110(3/4), 649–668.

NAS (1992). *Policy implications of greenhouse warming: mitigation, adaptation, and the science base*. National Academy of Sciences, Committee on Science, Engineering, and Public Policy. Washington, DC: National Academies Press.

Nichols, S. P., and Weldon, W. F. (1997). Professional responsibility: the role of the engineer in society. *Science and Engineering Ethics*, 3(3), 327–337.

Nightingale, P. (2004). Technological capabilities, invisible infrastructure and the un-social construction of predictability: the overlooked fixed costs of useful research. *Research Policy*, 33(9), 1259–1284.

Nowotny, H., Scott, P., and Gibbons, M. T. (2001). *Re-thinking science: knowledge and the public in an age of uncertainty*. Cambridge: Polity Press.

O'Donnell, E. (2013). Buffering the sun: David Keith and the question of climate engineering. *Harvard Magazine*, July/August. Available online at http://harvardmagazine.com/2013/07/buffering-the-sun (accessed 19 December 2013).

Petroski, H. (1996). *Invention by design: how engineers get from thought to thing*. Cambridge, MA: Harvard University Press.

——— (2004). *Pushing the limits: new adventures in engineering*. New York: Knopf.

Pierrehumbert, R. (2012). *Successful predictions*. Tyndall Lecture, presented at the American Geophysical Union Fall Meeting, San Francisco, 3–7 December 2012.

Pongratz, J., Lobell, D. B., Cao, L., and Caldeira, K. (2012). Crop yields in a geoengineered climate. *Nature Climate Change*, 2(2), 101–105.

Pope, F. D., Braesicke, P., Grainger, R. G., Kalberer, M., Watson, I. M., *et al.* (2012). Stratospheric aerosol particles and solar-radiation management. *Nature Climate Change*, 2(10), 713–719.

Rabinow, P., and Bennett, G. (2012). *Designing human practices: an experiment with synthetic biology*. Chicago: University of Chicago Press.

Rasch, P. J., Tilmes, S., Turco, R. P., Robock, A., Oman, L., *et al.* (2008). An overview of geoengineering of climate using stratospheric sulphate aerosols. *Philosophical Transactions of the Royal Society A: Mathematical, Physical & Engineering Sciences*, 366(1882), 4007–4037.

Rayner, S. (2000). Prediction and other approaches to climate change policy. In D. Sarewitz, R. A. Pielke, Jr, and R. Byerly, Jr (eds), *Prediction: science, decision making, and the future of Nature* (pp. 269–296). Covelo, CA: Island Press.

——— (2004). The novelty trap: why does institutional learning about new technologies seem so difficult? *Industry and Higher Education*, 18(6) 349–355.

Rier, D. A. (1999). The versatile 'caveat' section of an epidemiology paper: managing public and private risk. *Science Communication*, 21(1), 3–37.

Robock, A., Marquardt, A., Kravitz, B., and Stenchikov, G. (2009). Benefits, risks, and costs of stratospheric geoengineering. *Geophysical Research Letters*, 36(19), L19703.

Ryle, G. (1971). Knowing how and knowing that. In G. Ryle (ed.), *Collected papers*, Vol. 2: *Collected essays (1929–1968)* (pp. 212–225). London: Hutchinson.

Schyfter, P. (2013). Propellers and promoters: emerging engineering knowledge in aeronautics and synthetic biology. *Engineering Studies*, 5(1), 6–25.

Shackley, S., and Wynne, B. (1995). Global climate change: the mutual construction of an emergent science-policy domain. *Science and Public Policy*, 22(4), 218–230.

Staw, B. M. (1976). Knee-deep in the big muddy: a study of escalating commitment to a chosen course of action. *Organizational Behavior and Human Performance*, 16(1), 27–44.

Tkacik, D. (2011). Visiting professor discusses issues with geoengineering. *The Tartan*, 17 October. Available online at http://thetartan.org/2011/10/17/scitech/fts (accessed 29 July 2014).

Vincenti, W. G. (1990). *What engineers know and how they know it: analytical studies from aeronautical history*. Baltimore: Johns Hopkins University Press.

8 Reclaiming the experiment

Shared space

We have had more than a hundred years in which to come to terms with the motor car. The ways in which cars are regulated have evolved alongside the technology. In Chapter 2, I mentioned the *Traffic in Towns* report (Buchanan 1963), one of many attempts at a technology assessment. Since the publication of that report, planners and engineers in the UK and most other countries have worked on the assumption that the safest way to organise motorists and pedestrians was to separate them. Outside my own university in North London runs Euston Road, a busy dual carriageway decorated with the machinery of segregation – railings, warning signs, traffic lights, pedestrian crossings, curbs and underpasses. Outside Imperial College, on the other side of town, is an alternative arrangement. There, Exhibition Road has become an experiment in 'shared space'.

The idea of shared space comes from the Dutch traffic engineer Hans Monderman, whose aim is to create an architecture of responsibility. Monderman claims that 'we're losing our capacity for socially responsible behaviour. The greater the number of prescriptions, the more people's sense of personal responsibility dwindles' (quoted in Schulz 2006). His solution is to strip away the machinery and instead encourage drivers, pedestrians, cyclists and whomever else to pay attention to each other. Lines between roads and pavements are blurred, curbs become slopes and pedestrians are invited to cross wherever they choose. The uncertainty this creates leads to greater caution and, according to shared-space evangelists, improvements in safety, traffic flow, economic activity and the quality of public space. Shared space does not tell people what to do. Instead, it creates a set of conditions for rethinking relationships and responsibilities. As a framework for metagovernance, the idea has some important lessons for science and innovation.

Following growing recognition of the power of science and innovation to transform our lives in profound ways, we have seen the creation of regulatory machinery that seeks to protect our health, environment and rights. But much of this machinery contributes to the segregation of science from society. Risk assessments and ethics committees promise society safety from the downsides of innovation, but in doing so they exacerbate the illusion of the possibility of control. Geoengineering demonstrates the need to see governance as a shared space.[1]

There are plenty of arguments against the architecture of shared space. Opponents of the idea argue that by replacing rules with norms, shared space fails to protect vulnerable groups, such as blind pedestrians, and puts us all at risk from irresponsible drivers. One could make similar criticisms of a shared-space approach to science governance relating to power: vulnerable people still need to be protected, and rules are needed to ensure that bad intentions and bad actions don't endanger us all. The regulation of technology must take account of such asymmetries. But I would conclude that with the geoengineering ideas that have been the focus of this book, we are not there yet. Proposals for solar geoengineering are at an early stage, and we should resist the urge to attribute too much power to them.

What you think about shared-space roads is likely to depend on what you think about traffic and, in particular, its speed. Geoengineering researchers are quick to argue the need for more research. Whether the justification is the need to develop this technology in case of emergency or to understand it in the event of others' recklessness, there is a professed need for speed. It is hard to resist the tyranny of urgency and counter with a strong argument for slowness. Certainly there are some areas of innovation, directed at clear human needs, that demand strategic acceleration. But as I have argued in this book, for geoengineering, slow, thoughtful innovation is the order of the day (just as there is a case for 'slow food': Joly 2010). We should not pretend that we know how to proceed, what the relevant questions are, or whose expertise is most important. Perhaps we should follow the mantra of the shared-space movement that 'unsafe is safe'.

If we take current proposals at face value, geoengineering would seem to represent a radical new architecture of responsibility – the gradual technocratic ownership of the climate. But as I have described in this book, the reality of geoengineering research suggests a far more interesting picture of responsibility that should give us cause for optimism. Geoengineering researchers are sharing their space in novel and interesting ways, inviting non-scientists to help them navigate through their uncertainties. The concern is that as speed builds up, the instinct for segregation will take hold.

We shouldn't be scared of geoengineering, at least not yet. It is neither as exciting nor as terrifying as we have been led to believe, for the simple reason that it doesn't exist. The technologies of geoengineering, at least those that have generated the most discussion, remain imaginary. Our ignorance is vast, and scientific research, while generating knowledge, will produce further uncertainties. Projects such as SPICE that ask difficult questions about the means of geoengineering begin to reveal quite how hard it would be, while also generating important discussions about the ends of geoengineering research.

The big danger is not geoengineering but the fatalism, born of technological determinism, that is allowed to frame debates about emerging technologies. Part of my argument in this book is that geoengineering presents us – scientists, social scientists, stakeholders, members of the public – with an opportunity to imagine an alternative. The debate about geoengineering has provided a unique window onto a technology-in-the-making, a chance to see scientists agonising about an issue for which there are no right answers. One of the Royal Society working group members told me about the novelty of '*having conversations about values before the technology has come into being, and encountering all of the problems that are involved in having those conversations*'. During these conversations, scientists have acquired some of the language of social science, philosophy and law, and social scientists, philosophers and lawyers have begun moving closer to a constructive engagement with science. In this book, I have tried to report on some of the conversations I have staged, observed or been involved in. My hope is that they help chart a way forward not just for geoengineering research but for the governance of emerging technologies more generally.

The SPICE researchers began their project critical of the idea of geoengineering but with a conventional sense of their own responsibilities as scientists. The expansion and recasting of their highly publicised open-air experiment challenged them to rethink their responsibilities. The decision not to conduct the balloon experiment was an act of scientific responsibility. They were responding to concerns held within and outside the project about intellectual property and a lack of clear governance. But we cannot and should not rely on the good nature of publicly funded scientists to govern geoengineering research.

I have argued in this book that responsibility is about more than good intentions. The first point is that good intentions are not universally agreed upon. In my journey through the world of geoengineering research, I have met only well-intentioned scientists, but they would not share each other's visions for the Earth's future. Second, geoengineering is already being scientised, creating lock-ins and path dependencies that may prove irreversible. Momentum is gradually building up, even if many individual scientists would rather it didn't.

Geoengineering is not inevitable. Indeed, the closer one looks the less likely it appears. But as it becomes 'thinkable' as a topic of sensible scientific inquiry, we can assume there will be research findings and innovations that change relationships between people, technologies, the climate and politics, even if we can't predict what these changes will look like. There are various perspectives on the desirability of such changes. Some would argue that climate policy is urgently in need of disruption, given the apparent failure of our efforts so far. Others would say that now, more than ever, we need to stay the course. It is hard to imagine how geoengineering fits neatly into current global climate policy. As Mike Hulme has argued, geoengineering 'destabilises far more than it stabilises' (Hulme 2014, p. 55). Most would agree that this is something to which we should pay attention. So how should we proceed?

Guidelines for collective experimentation

Given the uncertainties of geoengineering, one could imagine two caricatured approaches. The first is technocracy, treating geoengineering as a technoscientific task. A technocratic approach would give engineers the responsibility to develop options, which would be assessed in conversation with scientists. The second approach is one of regulation. This approach would recognise that geoengineering was not an issue merely for science, but one about which there were legitimate concerns relating to ethics and risk. A regulatory approach would look to control the technologies proposed for geoengineering.

To their credit, there are few geoengineering scientists calling for an undiluted technocratic approach. The dominant view is that science should proceed, with substantial caution about the development of technological options until we know more and controls are in place for the use of any technologies that emerge. In this sense, the consensus is for a blend of both approaches. There has been much discussion among geoengineering researchers about where either is appropriate and how we might demarcate them. As I described in Chapter 5, there have been attempts to draw lines around and through the geoengineering issue to define an 'allowed zone' (Morgan *et al.* 2013) for scientific research. Scientists have argued about thresholds for the regulation of experiments, whether these are defined in terms of environmental impact or drawn at the door of a university laboratory. Some have attempted to draw disciplinary or technological lines to define what counts as geoengineering. Others have sought to reaffirm an informal taboo against talking about it at all. Outside geoengineering research, in the wider world of Earth system science, researchers have, through ideas such as the Anthropocene and planetary boundaries, sought to draw uncrossable lines of a different kind. As I have argued, this project of line drawing rests on assumptions not just of unfettered scientific freedom within social limits, but also of a Promethean technology. It contributes to the construction of geoengineering as a thing in the world, to be studied by natural scientists and with ethical implications to be speculated about. The more immediate concerns that come with recognising geoengineering as a work-in-progress are overlooked.

Both the technocratic and the regulatory approach are based on some flawed assumptions, not least the assumption of technological inevitability. The attempt to draw clear lines around science and policy will come unstuck because these things are coproduced (Jasanoff 2004). My suggested alternative, given that we have the time and space to deliberate, is to approach geoengineering as a collective experiment.

The idea of collective experimentation is both descriptive and prescriptive. As I described in Chapter 2, the experimentality of technologies that had been assumed to be predictable, controllable and well bounded has become more apparent over the last few decades. Although this diagnosis originally came from studies of technologies for energy and agriculture, our engagement with ubiquitous computing and social media has made the experiment more obvious. As users, we are still making sense of these technologies, and we do not presume that their creators can completely anticipate their future direction.

This description of technology prompts a new set of demands (Latour 2011): If there are experiments taking place, who is taking responsibility for them? What are the experimental ethics? What are the protocols? What are we learning? Understanding the experiment as a collective one means including institutions in the apparatus. We have seen with SPICE a version of a story that is familiar in science policy: scientists taking responsibility as individuals while the institutions that should be making key decisions fall away from view. Research funders, universities and the UK government have been largely silent on geoengineering, preferring to let individual scientists lead the public debate. The Royal Society's (2009) report is an important exception, which is why it remains a cornerstone for geoengineering debates. The Society's support for the Solar Radiation Management Governance Initiative reflects a sense of institutional responsibility, although this support was short-lived. The institutional apparatus of geoengineering has received very little attention, beyond half-hearted calls for some form of global body to take responsibility. Kaushik Sunder Rajan (2010) calls clinical trials the 'experimental machinery' of biocapital. We remain some way away from a geoengineering equivalent capable of building both relevant knowledge and public credibility. But we can start to consider what sorts of relationships should condition geoengineering research.

Given the uncertainties of emerging technologies, we can see immediate problems with a conventional regulatory mode demanding 'informed consent'. In the case of geoengineering, it is not clear how the planet's population could offer consent in a meaningful way (Szerszynski *et al.* 2013), nor is it clear what they would be consenting to. The information available is permanently partial. Sunder Rajan (2010) criticises clinical trials for seeing their subjects as 'merely risked' – in need of protection but no more. Making collective experimentation more responsible means finding ways for people to move from being subjects to being experimenters in themselves or, at least, being able to have a say about the direction of experimentation.

Imagining the public

Geoengineering reveals the opportunity for and challenges of upstream public engagement (see Wynne 2002; Wilsdon and Willis 2004). With geoengineering, we do not see the reaction experienced by those turning a critical eye to other emerging technologies, who are accused of being in some way 'anti-science' (see Stirling 2010). Unlike some more esoteric areas of science, the public nature of geoengineering research is inescapable. Even though the technology is unformed, scientists, social scientists and others are already imagining publics of various shapes and sizes. In my conversations with scientists, they invoked various 'you's, 'we's and 'they's as they grappled with the possible challenges of governing geoengineering. Many statements began with 'we' – 'we' might decide that 'we' need geoengineering; 'we' won't be able to control it; 'we' can't trust the models. But it was unclear who the 'we' was – experimentalists, scientists, experts, the global population, a specific group of citizens or someone else entirely.

Mike Hulme has argued that the quest for global temperature reduction suggests, or even demands, a 'global "we"' (Hulme 2014, pp. 54–55), a simplified, homogeneous mass capable of making decisions in the interests of all people and the planet. Scientists might legitimately claim that they are putting the complexities of governance to one side in order to investigate other aspects of geoengineering. But we have seen with past schemes and technologies the effects of globalising a problem in order to explore solutions. Sheila Jasanoff (2005) argues that such a mentality is at the root of controversies over genetically modified crops in developing countries. James Scott (1998) provides other cautionary tales of top-down modernism. In the case of geoengineering, the state currently being imagined is a global one. But the cautionary tales would suggest that given the unpredictability of global climates and people, technological enthusiasts may look to force predictability or 'legibility' (Scott 1998) onto people. Such a Taylorisation of the climate through scientific management seems incompatible with current models of democracy (Szerszynski *et al.* 2013).

As I described in Chapter 2, scientists' claims about the Anthropocene and planetary boundaries are making similar assumptions about global government and occasionally drifting into explicit arguments in favour of such top-down, anti-democratic arrangements. The view from space achieved by the satellites and models of climate science has unproblematically been turned into a call for governance, in effect, from space. Matters of governance and politics cannot simply be put to one side while research continues as though it were apolitical. Treating geoengineering as a collective experiment would demand that governance become part of the research question. I mentioned in Chapter 6 an early attempt to put people back into a climate modelling study by asking experts to take collective decisions about a simulated future climate. The project of truly democratic geoengineering research demands more such experiments.

Geoengineering has been the subject of a number of direct deliberative public engagement exercises, some of which I reviewed in Chapter 3. These have gone some way towards elucidating public concerns about geoengineering in general and the SPICE project in particular. One such exercise (Macnaghten and Szerszynski 2013) used insights from focus groups to anticipate the 'social constitution' of solar radiation management (SRM) technologies. Heywood and Rayner (2013) have criticised this work for prematurely identifying essential characteristics of a technology that remains imaginary. While I would agree that we should take seriously the technological uncertainties, we can still anticipate likely social and political reconfigurations. We should regard social and political ramifications as just as important as climatic ones and include them in our anticipation and experimentation. If exercises in public engagement are to make a difference to the geoengineering debate, we should look to draw out these sorts of ramifications, even if they are laden with uncertainty. The interpretation of public engagement findings by scientists and scientific institutions tends to downplay the profundity of public antipathy. As Carr and colleagues have concluded, public engagement asks more of scientists than most scientists realise (Carr *et al.* 2014).

Such processes should not be reduced to a social research exercise of merely finding out what the public thinks. Experiments in public engagement mark the beginning rather than the end of a much-needed public debate. There is still a tendency to pathologise the public, treating them as a problem to be 'solved' through conversation (Stilgoe *et al.* 2014). Rather than fixating on 'The Public', we should think about governance 'in public'. Initiatives such as the Oxford Principles for geoengineering (Rayner *et al.* 2013) are a useful start to this journey.

Crucially, however, the project of collective experimentation requires more than just superficial engagement with members of the public. It demands a reorientation in the process and representation of science itself.

Democratising uncertainty

Just as nanotechnology and genetics have changed the world, but not in the ways that their early proponents predicted (Fortun 2005), so geoengineering will change the way that we think about humans' relationship to the planet, but not in the ways currently being imagined. Rather than simply pointing out that we are ignorant about the future of geoengineering, we need to look at the ways in which uncertainty is being understood and controlled within and around the scientific community and identify opportunities for democratisation.

Uncertainty is clearly an important driving force for scientific research. Susan Leigh Star has argued that science is a process of controlling local uncertainties in the search for global certainties (Star 1985). But as the controversy over the SPICE project demonstrates, the relevant uncertainties are not immediately apparent. If a project dares to ask questions that are at odds with informally agreed norms, scientists outside the project can react just as defensively as NGOs. Most geoengineering researchers considered uncertainties about engineering less important, less interesting and more problematic than those that were amenable to climate modelling. This contributed, as I described in Chapter 3, to the construction of geoengineering as a set of fictional promises, rather than technical questions. Ethics and economics, in particular, have brought their own certainties to the debate and, in doing so, closed down discussions of uncertainties relating to the politics, as well as the technology, of geoengineering.

Collective experimentation means opening up the discussion of uncertainty as part of a democratisation of research agendas. Uncertainty and ambivalence should provide a common language for the democratisation of science. Public engagement is more constructive when the topic of conversation is uncertainty rather than evidence (Stilgoe 2007). But science habitually hides some uncertainties from view while formalising and presenting others, such as those surrounding climate model results (Jasanoff and Wynne 1998). Uncertainties are systematically treated as risks. The assumption is that the public demand certainty, and uncertainty, like experimentation, is imagined to be a scientific concern rather than a democratic one.

Uncertainty is a product of interdisciplinarity, as well as being its currency. New questions, challenges and surprises are posed by disciplinary clashes. Part of the value of interdisciplinarity is therefore that it disrupts the neat control of uncertainty. The same goes for public engagement, whether polite and planned or uninvited and noisy. According to Dominique Pestre, improvements to governance over the last few decades are not as rational as they first appear:

> Contrary to what managers, engineers, politicians and risk experts want to make us believe, it is the massive mobilization of the population, of dissident

> experts and of victims which have led ministerial departments, industrialists, safety committees and courts of justice to modify their attitudes.
>
> (Pestre 2013, p. 151)

And yet these dynamics are easy to forget. We convince ourselves that the process of social learning can be controlled by experts. We continue to imagine that technologies can be divorced from their governance. If we put politics and engineering back into the list of relevant uncertainties, we see the extent to which 'geoengineering' and 'governance' are coproduced.

Geoengineering as governance

I have in this book criticised the idea, which has already been allowed to take root, that geoengineering is a meaningful technology or set of technologies. The term is used as if it were a noun rather than a verb. If we instead think about geoengineering as a process, we invite new discussions of responsibility, ethics and experimentation. We are forced to admit that as with other emerging technologies, governance must take place in a state of uncertainty and ignorance. And the lines between science, technology and society are blurred.

Geoengineering, as an engineering challenge, would be a form of governance. The idea would be to govern the Earth's climate in a new way, to take control of something that has until now been considered largely accidental. Viewed in this way, the challenges of control look vast, from the perspective of both politics and engineering (see Stirling 2014 for a fuller discussion). Geoengineering as a noun looks 'easy' (Keith 2013). As a verb, it looks all but impossible. Brad Allenby and Dan Sarewitz (2011) claim that 'we've made a world we cannot control'. Most people, assuming a strict sense of 'control', would agree. But while recognising the limits of our ability to control, we could also strive for greater controllability as a criterion for metagovernance. As I describe in Chapter 2, a key part of the project of responsible innovation is to make it possible to take responsibility and draw lines of accountability in the future. The way forward is better described as 'care' than as 'control' (Stilgoe *et al.* 2013; Stirling 2014). Just as perfect climate 'steering' (Hale 2013, p. 201) will prove impossible, so we should not pretend that geoengineering research can be steered in a single, responsible direction. But research trajectories may be modulated in more responsible ways (see Fisher *et al.* 2006).

In his polemic against stratospheric particle injection, Mike Hulme (2014) provides a strong argument against the technological optimism of David Keith (2013). The power of his argument lies in his reintroduction of sociopolitical concerns. The technology, he argues, looks ungovernable. Similarly, Richard Owen (2014) asks why SRM has been legitimised as an object of governance at all. I agree that we should hold on to the possibility of complete rejection of particular technological options. But a closer look at the world of geoengineering research shows how unlikely the technology currently being imagined is to materialise. In ruling it out, we pretend to know what 'it' is. We may be able to prohibit a particular proposal, but this may deflect attention away from, and so ease the progress of, others. Stratospheric particle injection is not even one idea. It is characterised by multiple technologies and multiple intents. Our only option is an approach of continued vigilance, collective experimentation and scrutiny of technologies as they begin to stabilise.

Experimentation is a way of destabilising current understandings by introducing surprises. Collective experimentation involves broadening the scope of these surprises in conversation between disciplines and with others outside the research community. This is not about disempowering scientists, but rather allowing them to express their own ambivalence and articulate their responsibilities in public to prevent overly narrow understandings of geoengineering from getting too comfortable.

At the time of writing, some geoengineering researchers are starting to lobby for research funding in the open scientific press as well as behind closed doors in conversations with funders whom they regard as overly cautious. The debate has been cast such that sensible people are imagined to be anti-geoengineering but pro-geoengineering research. 'Research' is a one-dimensional thing in this view – a choice between knowledge and ignorance. 'Given this choice', David Keith argues, 'I choose research' (Keith 2013, p. 13). If this is indeed the choice, it is utterly disempowering for anyone except a geoengineering researcher. Thankfully, the potential for democratic governance is broader than this. We should instead be asking what sorts of research should be supported and how.

Collective experimentation sees research as conditional and relational. Following the arguments of this book, I would argue that geoengineering research should proceed according to some important principles and practicalities. Making geoengineering more anticipatory, more inclusive, more reflexive and more responsive (see Chapter 2) means creating new research relationships. SPICE's sister project, the Integrated Assessment of Geoengineering Proposals, was a step in this direction. But we should not underestimate the work required in the coming together of different disciplines and their engagement with unfamiliar perspectives. The social work of geoengineering research may be as time-consuming as the scientific work. Collective experimentation is also experimentation in the organisation of science and its institutions. The shared space of geoengineering research is an ideal location in which to rethink the relationship between science, politics and the public.

Note

1 This analogy owes credit to Bronislaw Szerszynski. It began life as a conversation with Bron in an Oxford traffic jam, on the way home from a geoengineering conference.

References

Allenby, B. R., and Sarewitz, D. (2011). We have made a world we cannot control. *New Scientist*, 210(2812), 28–29.

Buchanan, C. (1963). *Traffic in towns: the Buchanan report*. London: HMSO.

Carr, W., Yung, L., and Preston, C. (2014). Swimming upstream: engaging the American public early on climate engineering. *Bulletin of the Atomic Scientists*, 70(3), 38–48.

Fisher, E., Mahajan, R. L., and Mitcham, C. (2006). Midstream modulation of technology: governance from within. *Bulletin of Science, Technology & Society*, 26(6), 485–496.
Fortun, M. (2005). For an ethics of promising, or: a few kind words about James Watson. *New Genetics and Society*, 24(2), 157–174.
Hale, B. (2013). Remediation vs. steering: an act-description approach to approving and funding geoengineering research. In J. Basl and R. L. Sandler (eds), *Designer biology: the ethics of intensively engineering biological and ecological systems* (pp. 195–218). New York: Lexington Books.
Heywood, C., and Rayner, S. (2013). *A curious asymmetry: social science expertise and geoengineering*. CGG Working Paper No. 7. Oxford: Climate Geoengineering Governance, Institute for Science, Innovation and Society.
Hulme, M. (2014). *Can science fix climate change? A case against climate engineering*. Cambridge: Polity Press.
Jasanoff, S. (ed.) (2004). *States of knowledge: the co-production of science and the social order*. London: Routledge.
——— (2005). 'Let them eat cake': GM foods and the democratic imagination. In M. Leach, I. Scoones and B. Wynne (eds), *Science and citizens: globalization and the challenge of engagement* (pp. 183–198). London: Zed Books.
Jasanoff, S., and Wynne, B. (1998). Science and decision-making. In S. Rayner and E. Malone (eds), *Human choice & climate change*, Vol. 1: *The societal framework* (pp. 1–87). Columbus, OH: Battelle Press.
Joly, P. B. (2010). On the economics of techno-scientific promises. In M. Akrich, Y. Barthe, F. Muniesa and P. Mustar (eds), *Débordements : mélanges offerts à Michel Callon* (pp. 203–222). Paris: Presses des Mines.
Keith, D. (2013). *A case for climate engineering*. Cambridge, MA: MIT Press.
Latour, B. (2011). From multiculturalism to multinaturalism: what rules of method for the new socio-scientific experiments? *Nature and Culture*, 6(1), 1–17.
Macnaghten, P., and Szerszynski, B. (2013). Living the global social experiment: an analysis of public discourse on solar radiation management and its implications for governance. *Global Environmental Change*, 23(2), 465–474.
Morgan, M. G., Nordhaus, R. R., and Gottlieb, P. (2013). Needed: research guidelines for solar radiation management. *Issues in Science and Technology*, 29(3), 37–44.
Owen, R. (2014). Solar radiation management and the governance of hubris. In R. E. Hester and R. M. Harrison (eds), *Issues in environmental science and technology*, Vol. 38: *Geoengineering of the climate system* (pp. 211–247). Cambridge: Royal Society of Chemistry.
Pestre, D. (2013). À contre-science : politiques et savoirs des sociétés contemporaines (transl H. Nowotny). Lecture presented at the London School of Economics, 15 October. Paris: Éditions du Seuil.
Rayner, S., Heyward, C., Kruger, T., Pidgeon, N., Redgwell, C., and Savulescu, J. (2013). The Oxford Principles. *Climatic Change*, 121(3), 499–512.
Royal Society (2009). *Geoengineering the climate: science, governance and uncertainty*. London: Royal Society.
Schulz, M. (2006). Controlled chaos: European cities do away with traffic signs. *Der Spiegel*, 16 November. Available online at http://www.spiegel.de/international/spiegel/controlled-chaos-european-cities-do-away-with-traffic-signs-a-448747.html (accessed 29 July 2014).
Scott, J. C. (1998). *Seeing like a state: how certain schemes to improve the human condition have failed*. New Haven, CT: Yale University Press.

Star, S. L. (1985). Scientific work and uncertainty. *Social Studies of Science*, 15(3), 391–427.

Stilgoe, J. (2007). *Nanodialogues: experiments in public engagement with science*. London. Demos.

Stilgoe, J., Lock, S. J., and Wilsdon, J. (2014). Why should we promote public engagement with science? *Public Understanding of Science*, 23(1), 4–15.

Stilgoe, J., Owen, R., and Macnaghten, P. (2013). Developing a framework for responsible innovation. *Research Policy*, 42(9), 1568–1580.

Stirling, A. (2010). Keep it complex. *Nature*, 468(7327), 1029–1031.

——— (2014). *Emancipating transformations: from controlling 'the transition' to culturing plural radical progress*. CGG Working Paper No. 12. Oxford: Climate Geoengineering Governance, Institute for Science, Innovation and Society. Available online at http://www.geoengineering-governance-research.org/perch/resources/workingpaper12stirlingemancipatingtransformations.pdf (accessed 18 July 2014).

Sunder Rajan, K. (2010). The experimental machinery of global clinical trials: case studies from India. In A. Ong and N. N. Chen (eds), *Asian biotech: ethics and communities of fate* (pp. 55–80). Durham, NC: Duke University Press.

Szerszynski, B., Kearnes, M., Macnaghten, P., Owen, R., and Stilgoe, J. (2013). Why solar radiation management geoengineering and democracy won't mix. *Environment and Planning A*, 45(12), 2809–2816.

Wilsdon, J., and Willis, R. (2004). *See-through science: why public engagement needs to move upstream*. London: Demos.

Wynne, B. (2002). Risk and environment as legitimatory discourses of technology: reflexivity inside out? *Current Sociology*, 50(3), 459–477.

Index